T0331752

Granular Materials at the Meso-scale

Discrete Granular Mechanics Set

coordinated by
Félix Darve

Granular Materials at the Meso-scale

Towards a Change of Scale Approach

Bernard Cambou
Hélène Magoariec
Ngoc-Son Nguyen

ELSEVIER

First published 2016 in Great Britain and the United States by ISTE Press Ltd and Elsevier Ltd

ISTE Press Ltd
27-37 St George's Road
London SW19 4EU
UK

www.iste.co.uk

Elsevier Ltd
The Boulevard, Langford Lane
Kidlington, Oxford, OX5 1GB
UK

www.elsevier.com

Notices

Knowledge and best practice in this field are constantly changing. As new research and experience broaden our understanding, changes in research methods, professional practices, or medical treatment may become necessary.

Practitioners and researchers must always rely on their own experience and knowledge in evaluating and using any information, methods, compounds, or experiments described herein. In using such information or methods they should be mindful of their own safety and the safety of others, including parties for whom they have a professional responsibility.

To the fullest extent of the law, neither the Publisher nor the authors, contributors, or editors, assume any liability for any injury and/or damage to persons or property as a matter of products liability, negligence or otherwise, or from any use or operation of any methods, products, instructions, or ideas contained in the material herein.

For information on all our publications visit our website at http://store.elsevier.com/

British Library Cataloguing-in-Publication Data
A CIP record for this book is available from the British Library
Library of Congress Cataloging in Publication Data
A catalog record for this book is available from the Library of Congress
ISBN 978-1-78548-065-2

Printed and bound in the UK and US

Contents

Foreword

Molecular dynamics is recognized as a powerful method in modern computational physics. This method is essentially based on a factual observation: the apparent strong complexity and extreme variety of natural phenomena are not due to the intrinsic complexity of the element laws but due to the very large number of basic elements in interaction through, in fact, simple laws. This is particularly true for granular materials in which a single intergranular friction coefficient between rigid grains is enough to simulate, at a macroscopic scale, the very intricate behavior of sand with a MohrCoulomb plasticity criterion, a dilatant behavior under shearing, non-associate plastic strains, etc. and, in fine, an incrementally nonlinear constitutive relation. Passing in a natural way from the grain scale to the sample scale, the discrete element method (DEM) is precisely able to bridge the gap between micro- and macro-scales in a very realistic way, as it is today verified in many mechanics labs.

Thus, DEM is today in an impetuous development in geomechanics and in the other scientific and technical fields related to grain manipulation. Here lies the basic reason for this new set of books called "Discrete Granular Mechanics", in which not only numerical questions are considered but also experimental, theoretical and analytical aspects in relation to the discrete nature of granular media. Indeed, from an experimental point of view, computational tomography for example is giving rise today to the description of all the translations and rotations of a few thousand grains inside a given sample and to the identification of the formation of mesostructures such as force chains and force loops. With respect to theoretical aspects, DEM is also confirming, informing or at least precising some theoretical clues such as the questions of failure modes, of the expression of stresses inside a partially saturated medium and of the mechanisms involved in granular avalanches. Effectively, this set has been planned to cover all the experimental, theoretical and numerical approaches related to discrete granular mechanics.

The observations show undoubtedly that granular materials have a double nature, that is continuous and discrete. Indeed, roughly speaking, these media respect the matter continuity at a macroscopic scale, whereas they are essentially discrete at the granular microscopic scale. However, it appears that, even at the macroscopic scale, the discrete aspect is still present. An emblematic example is constituted by the question of shear band thickness. In the framework of continuum mechanics, it is well recognized that this thickness can be obtained only by introducing a so-called "internal length" through "enriched" continua. However, this internal length seems to be not intrinsic and to constitute a kind a constitutive relation by itself. Probably, it is because to consider the discrete nature of the medium by a simple scalar is oversimplifying reality. However, in a DEM modeling, this thickness is obtained in a natural way without any ad hoc assumption. Another point, whose proper description was indomitable in a continuum mechanics approach, is the post-failure behavior. The finite element method, which is essentially based on the inversion of a stiffness matrix, which is becoming singular at a failure state, meets some numerical difficulties to go beyond a failure state. Here also, it appears that DEM is able to simulate fragile, ductile, localized or diffuse failure modes in a direct and realistic way even in some extreme cases such as fragmentation rupture.

The main limitation of DEM is probably linked today to the limited number of grains or particles, which can be considered in relation to an acceptable computation time. Thus, the simulation of boundary value problems stays, in fact, bounded by more or less heuristic cases. So, the current computations in labs involve at best a few hundred thousand grains and, for specific problems, a few million. Let us note however that the parallelization of DEM codes has given rise to some computations involving 10 billion grains, thus opening widely the field of applications for the future.

In addition, this set of books will also present the recent developments occurring in micromechanics, applied to granular assemblies. The classical schemes consider a representative element volume. These schemes are proposing to go from the macro-strain to the displacement field by a localization operator, then the local intergranular law relates the incremental force field to this incremental displacement field, and eventually a homogenization operator deduces the macro-stress tensor from this force field. The other possibility is to pass from the macro-stress to the macro-strain by considering a reverse path. So, some macroscopic constitutive relations can be established, which properly consider an intergranular incremental law. The greatest advantage of these micromechanical relations is probably to consider only a few material parameters, each one with a clear physical meaning.

This set of around 20 books has been envisaged as an overview toward all the promising future developments mentioned earlier.

Félix DARVE
July 2015

Acknowledgments

Particular thanks to Marie Chaze and Sinh-Khoa Nguyen, who allowed a part of their personal research work to be included in this book.

We are grateful to our colleagues Alexandre Danescu and Eric Vincens, who have been involved in the development of certain parts of the research presented in this book.

Preface

In the first part of the 20th century, more specifically after World War II, all developed countries put in place ambitious infrastructure plans in the domains of transport (roads, railways, bridges, tunnels, etc.), energy and water supply (dams, dykes, etc.). Building these infrastructures has required development of accurate technical tools, in particular, for the modeling of soil behavior, and more specifically of granular soils. By the onset of this century, the design of these works was based on classical continuum mechanics; only elasticity was available for the forecast of small strains in such a material, and perfect plasticity for the analysis of potential failures. These two crude approximations associated with the personal experiences of engineers make major development of these infrastructures possible, which has contributed to significant social and economical progress throughout this century.

Since 1960 a new technical tool appeared in different domains of engineering, in particular, in civil engineering: the numerical method called the *Finite Element Method* (FEM). This tool, developed in the framework of continuum mechanics, allows complex geometrical works to be analyzed and nonlinear constitutive laws to be taken into account in the civil engineering design. This method was initially considered as a tool able to solve any kind of civil engineering problem, in particular in soil mechanics. However, engineers discovered very soon that the use of this tool was not so easy. The definition of accurate nonlinear constitutive laws was difficult for complex materials such as granular soils, and the definition of numerical values of numerous parameters of these laws was often a complex problem. Thus, many open problems persisted for easy and efficient use of this tool.

Since 1979 with the pioneering work of P. Cundal [CUN 79], a new numerical method called the *Discrete Element Method* (DEM) has been developed. This method is very different from the FEM as it takes into account the discrete nature of granular soil, considering the movement of each grain and its interactions with all other grains. This method is well known today, and has been developed in research laboratories to analyze very accurately the mechanical behavior of granular material in relation with its internal texture. This approach has allowed a better understanding of the evolution of the texture of granular materials along a given loading path as well as some improvements of the constitutive laws used for the modeling of granular materials. This method is now also used for the modeling of actual works, but it remains complex because of the high number of grains to be considered on actual works, whereas simulations can only consider a certain number of these grains.

Therefore, four approaches can be used for the design of works including granular materials, in particular granular soils:

– analytical approach using elasticity for small strain and perfect plasticity for the analysis of failure. This approach can only give a first approximation of simple problems;

– numerical modeling using the FEM including a constitutive law. This approach can give interesting results, although it requires improvement of the used constitutive law and to develop experimental tests to define the values of its parameters;

– numerical modeling using the DEM. This approach can be used for works built with large blocks, such as retaining wall or rockfill dams, but cannot be used for works with a very large number of grains;

– numerical modeling using a multi-scale approach based on DEM at the Representative Elementary Volume (REV) scale and on FEM at the work scale. This approach seems to be of great interest and is currently in development in research teams.

This book is dedicated to research which has been developed in the framework of the second item written above. In the first part of this book accurate and innovative analysis performed at an intermediate scale, the *meso-scale*, of the evolution of the internal texture of granular material is presented. This can be of great interest to improve phenomenological constitutive laws for granular materials. It can help to develop these constitutive laws to include a better description of the internal state along a given loading. The last part of this book is dedicated to an accurate development allowing us to build a complete constitutive law based on a change of scale approach, using a meso-scale.

Symbols

Mathematical symbols

δ_{ij} Kronecker symbol:

$$\delta_{ij} = \begin{cases} 0 \text{ if } i \neq j, \\ 1 \text{ if } i = j. \end{cases}$$

ϵ_{ijk} Permutation symbol:

$$\epsilon_{ijk} = \begin{cases} 0 \quad \text{for } i = j, \; j = k, \text{ or } k = i, \\ +1 \text{ for } (i,j,k) \in \{(1,2,3), \; (2,3,1), \; (3,1,2)\}, \\ -1 \text{ for } (i,j,k) \in \{(1,3,2), \; (3,2,1), \; (2,1,3)\}. \end{cases}$$

I Identity matrix

$\cdot$ Single dot product of two vectors: $c = a \cdot b$ and $c = a_i b_i$

$\cdot$ Dot product of a matrix and a vector: $b = A.a$ and $b_i = A_{ij} b_j$

$\wedge$ Cross product of two vectors: $c = a \wedge b$ and $c_i = \epsilon_{ijk} a_j b_k$

$\otimes$ Tensor product of two vectors: $C = a \otimes b$ and $C_{ij} = a_i b_j$

$\otimes^s$ Symmetric tensor product: $C = a \otimes^s b$ and $C_{ij} = (a_i b_j + a_j b_i)/2$

:	Double dot product of two tensors: $A : B = A_{ij}B_{ij}$
$\|.\|$	2D-norm of a vector or a matrix: $\|a\| = \sqrt{a_i a_i}$ and $\|A\| = \sqrt{A_{ij}A_{ij}}$
$I_1(A)$	First invariant of a 2-by-2 matrix A: $I_1(A) = A_{kk}$
$I_2(A)$	Second invariant of a 2-by-2 matrix A: $I_2(A) = \sqrt{A_{ij}A_{ij}/2}$
$\mathrm{Dev}(A)$	Deviatoric part of a 2-by-2 matrix A: $\mathrm{Dev}(A) = A - \dfrac{I_1(A)}{2}\mathbf{I}$

Stress and strain defined at different scales for 2D materials

σ	Stress tensor defined over a given volume
σ_1, σ_2	Major and minor principal stresses defined for a given volume, respectively
p	Mean stress defined for a given volume: $$p = \frac{I_1(\sigma)}{2} = \frac{\sigma_1 + \sigma_2}{2}$$
q	Deviatoric stress defined for a given volume: $$q = I_2(\mathrm{Dev}(\sigma)) = \frac{\sigma_1 - \sigma_2}{2}$$
a	Displacement gradient tensor for a given volume
ε	Strain tensor defined for a given volume
$\varepsilon_1, \varepsilon_2$	Major and minor principal strains defined for a given volume, respectively
ε_v	Volumetric strain defined for a given volume: $$\varepsilon_v = I_1(\varepsilon) = \varepsilon_1 + \varepsilon_2 = \varepsilon_{kk}$$
ε_d	Deviatoric strain defined for a given volume: $$\varepsilon_d = 2I_2(\mathrm{Dev}(\varepsilon)) = \varepsilon_1 - \varepsilon_2$$

Variables defined at the sample level

Σ	Macroscopic stress tensor
$\hat{\Sigma}$	Modified macroscopic stress tensor used to define a static localization operator
Σ_1, Σ_2	Macroscopic major and minor principal stresses, respectively
Q	Macroscopic deviatoric stress
P	Macroscopic mean stress
Q/P	Macroscopic stress ratio
E	Macroscopic strain tensor
E_v	Macroscopic volumetric strain
δA	Increment of the macroscopic displacement gradient tensor
δE	Increment of the macroscopic strain tensor
Φ	Porosity of a sample
e	Void ratio of a sample
V	Total volume of a sample

Local variables defined for contacts and internal variables

N_c	Number of contacts
f^k	Force at a contact k
f_n^k	Normal force at a contact k
f_s^k	Tangential force at a contact k
n^k	Normal vector at a contact k
l^k	Branch vector at a contact k
l'^k	Branch vector with or without contact
n_b^k	Unit vector directed along a branch k
b^k	Vector outward normal to the branch vector l^k
c^k	Relative displacement at a contact k
c_n^k	Normal relative displacement at a contact k
c_s^k	Tangential relative displacement at a contact k
δc^k	Increment of relative displacement at a contact k
δu^k	Increment of relative displacement between two particle centers at a contact k

$\dot{c}^k$ Relative velocity at a contact k
$\dot{c}_n^k$ Normal relative velocity at a contact k
$\dot{c}_s^k$ Tangential relative velocity at a contact k
H Contact fabric tensor
H' Branch fabric tensor
H'' Combined fabric tensor
H''' Orientation fabric tensor
H^* Modified contact fabric tensor
$\hat{H}$ Modified branch fabric tensor
Γ Coordination number
Γ^* Modified coordination number
$\hat{R}$ Mean redundancy number
R Mean valence over all meso-domain in a sample

Local variables defined for particles

N_g Number of particles
G^g Center of each particle
x^g Position of the center of each particle
b^g Vector defining the orientation of each particle
δu^g Increment of displacement of each particle center
ω^g Rotation of each particle
f^g Resultant of contact forces applied on each particle in a meso-domain by its external neighboring particles
N_c^g Number of contacts on each particle
D^g Dirichlet cell associated to each particle
V_D^g Volume of each Dirichlet cell
ΔV_D^g Volume fraction of Dirichlet cell, D^g, inside a meso-domain
σ^g Stress tensor defined for each Dirichlet cell

Variables defined for triangular elements

N_e Total number of triangular elements
Ve Volume of each triangular element
δa^e Displacement gradient tensor for each triangular element

$\delta\varepsilon^e$ Strain tensor for each triangular element

δu^{wv} Relative displacement of a vertex w to a vertex v

x^{wv} Vector joining a vertex w to a vertex v

y^{wv} Dual vector of x^{wv} used to define the displacement gradient tensor, δa^e

b^{wv} Vector obtained by rotating outward the vector x^{wv} by 90°

δu^s Relative displacement between two vertices on a side s

b^s Vector identical to b^{wv} with w and v being two vertices on side s

Q^3 Kuhn's matrix defined for a triangular element

Variables defined for meso-domains

V^m Volume of a meso-domain

N_m Number of meso-domains

σ^m Stress tensor defined for a meso-domain

p^m Mean stress defined for a meso-domain

q^m Deviatoric stress defined for a meso-domain

δa^m Increment of displacement gradient tensor of a meso-domain

$\delta\varepsilon^m$ Increment of strain tensor of a meso-domain

Q^{r^m} Kuhn's matrix defined for a meso-domain with valence r^m

ϕ^m Porosity of a meso-domain

r^m Valence of a meso-domain

L^m Loop tensor of a meso-domain

D^m Deviatoric part of the loop tensor of a meso-domain

β^m Elongation degree of a meso-domain

m^m Elongation direction of a meso-domain

θ^m Orientation angle of a meso-domain

$\hat{V}^m$ Volume of a meso-domain used to define meso-stress tensor

Variables defined for phases

$\mathcal{W}$ $(\mathcal{S})$ Class of weakly (strongly) elongated meso-domains

· $\mathcal{W}1'$ ($\mathcal{S}1'$) Phase of meso-domains weakly (strongly) elongated and oriented in the compression direction

$\mathcal{W}2'$ ($\mathcal{S}2'$) Phase of meso-domains weakly (strongly) elongated and oriented in the extension direction

$\mathcal{W}0'$ ($\mathcal{S}0'$) Phase of meso-domains weakly (strongly) elongated and oriented obliquely

σ^p Stress tensor defined for a phase

σ_1^p, σ_2^p Major and minor principal stresses for each phase

q^p Deviatoric stress defined for a phase

p^p Mean stress defined for a phase

r_σ^p Stress ratio for a phase

$\delta\varepsilon^p$ Strain tensor defined for each phase

$\delta\varepsilon_1^p$, $\delta\varepsilon_2^p$ Major and minor principal strains defined for each phase

$\delta\varepsilon_v^p$ Volumetric strain defined for each phase

$\delta\varepsilon_d^p$ Deviatoric strain defined for each phase

r_ε^p Strain ratio for a phase

$\boldsymbol{T}^p$ Texture tensor of a phase

T_d^p Anisotropy measure of a phase

P_v^p Volume fraction of a phase

r^p Valence of a phase

ϕ^p Porosity of a phase

Variables used in the CJS model for each phase

$\dot{\varepsilon}^{p,\text{tot}}$ Total incremental strain tensor

$\dot{\varepsilon}^{p,\text{el}}$ Incremental elastic strain tensor

$\dot{\varepsilon}^{p,\text{pl}}$ Incremental plastic strain tensor

$\boldsymbol{X}^p$ Anisotropy tensor

R_{char}^p Characteristic radius

γ^p Dilatancy parameter

β^p Parameter characterizing the plastic volume change

R_{fail}^p Failure radius

a^p Variable ruling velocity of the kinematic hardening

b^p, c^p Plastic hardening parameters

R_e Elastic radius of the yielding surface

E Young's modulus

ν Poisson's ratio

$P^p_{v\,\text{crit}}$ Volume fraction of a phase at the critical state

$r^p_{\varepsilon\,\text{crit}}$ Strain ratio at the critical state

Introduction

Granular materials are composed of grains in contact. These materials are, therefore, discontinuous and highly heterogeneous. Two or three phases can generally be defined in these materials (e.g. air and solid or air, water and solid). Due to this discrete nature, the mechanical behavior of granular materials is quite complex and phenomenological models available in the literature are not able to make accurate forecasts for complex loadings, such as cyclic loadings or loadings with rotation of principal stress directions. Furthermore, these phenomenological models consider a large set of parameters which are often difficult to identify from experimental tests and which may have unclear physical meanings.

Macroscopic mechanical properties of these materials are obviously linked to the local texture and to mechanical properties of each component (grains and voids) as well as to their interactions. It is, then, highly interesting to define the behavior of such discontinuous and heterogeneous materials at the macroscopic scale from characteristics defined at a local scale. This kind of approach has been widely developed for heterogeneous continua (fluids or solids) and is known as the *change of scale approach*. The aim of this method consists of building constitutive models from local characteristics and from some measures of the material fabric [WAL 87, JEN 91, CAM 95, EME 96, SAB 97, LIA 97, GOL 02, CHA 05]. This kind of approach is highly different from the phenomenological one, in which the constitutive model is derived from the general laws of thermodynamics considered in some particular cases such as elasticity, elasto-plasticity, etc. The macroscopic laws derived from local mechanical and structural information are, thus, expected to be more efficient than the aforementioned phenomenological laws.

In the last few years, different multi-scale approaches have been proposed in the literature for granular materials. Several papers [WAL 87, CHA 90, CAM 95, EME 96] consider a local scale, called *the micro-scale*, defined at the level of contacts between particles. At this scale, the local static variables (inter-particle contact forces) can easily be related to the static variable (stress tensor) at the macro-scale defined on the representative elementary volume (REV). However, it is more difficult to relate the kinematic variables defined at contacts (relative displacements at contacts) to the macroscopic kinematic variable (the strain tensor) directly, as proved by several authors [CAM 00, BAG 06]. This approach is presented in a synthetic way in Chapter 1. In another approach [CHA 05], the authors considered, at the micro-scale, a local behavior law similar to classical global phenomenological behavior laws for granular materials. Thus, this approach can be considered as a mixture between a multi-scale approach and a phenomenological approach. It is interesting because it allows microscopic properties to be added to a phenomenological approach. However, it presents some drawbacks of the usual phenomenological approaches mentioned above.

To overcome these difficulties when considering the micro-scale, some multi-scale approaches have introduced or suggested an intermediate scale between the micro-scale and macro-scale, which is defined at the level of clusters of particles [SAT 92, KRU 96, LIA 97, SAB 97, KUH 97, KUH 99, DUR 10a, DUR 10b, NIC 11, ZHU 16]. Such an intermediate local scale is expected to allow the local structure of granular materials to be described more precisely and the local static and kinematic variables to be more consistently defined. In several previously published papers [NGU 09b, NGU 12, NGU 14, CHA 14, CAM 14, NGU 15], we have analyzed the behavior of 2D granular materials at such a local scale based on different numerical simulations of 2D granular samples subjected to different kinds of loading. This book is essentially based on the results presented in these papers. In Chapter 2 this intermediate local scale, called the *meso-scale*, is defined for 2D granular materials using the particle graph proposed by [SAT 92]. A 2D granular material is entirely subdivided into *meso-domains*, each of which is enclosed by branches which join the centers of particles in contact. Each meso-domain is, thus, a closed loop of particles in contact.

In Chapter 3, the meso-structure of 2D granular assemblies is described and meso-stress and meso-strain are defined and analyzed consistently. It is shown that meso-stress and meso-strain fields are significantly structured, i.e. meso-stress and meso-strain are highly governed by the meso-domain elongation degree and orientation with respect to the principal compression direction. We then define six *phases*, which are sets of meso-domains with similar elongation degrees and orientations. In Chapter 4, the evolution of different variables defined at the meso-scale (texture, strain, stress) are then analyzed for the six defined phases and the mechanical behavior of these phases is analyzed in relation with the evolution of their internal texture. In Chapter 5, we propose a modeling of the mechanical behavior of the six considered phases by using an elastic-hardening-plastic model. A set of constitutive parameters is identified for each phase on a usual loading path. It is shown that these constitutive parameters can be correlated with two parameters: the initial anisotropy used in the elastic-plastic model considered for each phase and the orientation of each phase with respect to the loading direction. This chapter, finally, deals with a validation of the proposed model on a loading path different from the one used for the parameter identification.

Thus, we evidence that the work presented in this book is a demonstration of the feasibility of a change of scale approach using a meso-scale for granular materials. We show that a macroscopic constitutive model can be built on the basis of the modeling of the behavior of a finite number of phases which are defined by two texture parameters (elongation degree and orientation with respect to the loading direction). For a given loading, the macroscopic response can be built considering the behavior of each phase, their orientation with respect to the considered loading and the evolution of their volumetric amount. This approach can be applied to actual 3D granular materials in a similar way, as proposed by several authors concerning mechanical models for fracture materials [BAZ 90]. More research is needed to fulfill this goal.

Finally, we would like to draw the attention of readers to two fundamental points for the understanding of this book: the precise description of the different scales considered in this book and the conventions related to the loading directions.

I.1. The different scales considered

– *The microscopic scale:* this is the scale of particles and of contacts between particles. Numerical simulations using the Discrete Element Method (DEM) are based on mechanical laws written at this scale. Variables considered at this scale are positions and velocities of particle centers, rotations of particles, positions of contact points and contact forces.

– *The mesoscopic scale*: this is the scale of local arrays defined as closed loops of particles in contact, here called *meso-domains*. Variables considered at this scale are texture variables and local strain and stress tensors. These variables are defined from variables defined at the microscopic scale.

– *The aggregate mesoscopic scale*: this is the scale of the so-called *phases*, defined as sets of meso-domains sharing common texture characteristics (elongation degree and orientation). Variables considered at this scale are the volume fraction occupied by each phase, anisotropy tensor and stress and strain tensors defined for each phase. These variables are defined from variables defined at the mesoscopic scale.

– *The macroscopic scale*: this is the scale of the representative elementary volume (REV). Phenomenological constitutive laws for a given granular material are usually defined at this scale. Variables considered at this scale are stress and strain tensors besides some internal state variables. These variables can be obtained either from the boundary conditions of the considered REV or from variables defined at the aggregate mesoscopic scale.

I.1.1. *Conventions related to the loading directions*

The usual convention for granular material mechanics is used in this book: compressive stress and strain are positive. At the macroscopic scale we define stress and strain tensors in the fixed directions 1 and 2, which do not change. Nevertheless, along a given loading path, the role of axes 1 and 2 can change. To define what we call *the direction of loading*, we consider the increment of the deviatoric strain tensor:

$$\dot{\varepsilon}_{ij}^{d} = \dot{\varepsilon}_{ij} - \frac{1}{2}\dot{\varepsilon}_{kk}\delta_{ij}, \qquad\qquad [\text{I.1}]$$

where δ_{ij} is Kronecker symbol. Direction $1'$ (respectively $2'$) is the major (respectively minor) principal direction of the deviatoric strain rate, $\dot{\varepsilon}^d$. The major (respectively minor) principal value, $\dot{\varepsilon}^d_{1'}$ (respectively $\dot{\varepsilon}^d_{2'}$), of the deviatoric strain rate, $\dot{\varepsilon}^d$, is positive (respectively negative) as illustrated in Figure I.1. For instance, for a compressive loading in direction 1 we have $1' \equiv 1$ and $2' \equiv 2$, and for a compressive unloading in direction 1 we have $1' \equiv 2$ and $2' \equiv 1$.

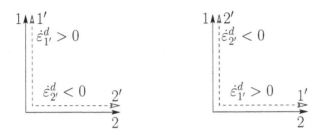

a) For the loading stage   b) For the unloading stage

Figure I.1. *The two systems of axes, $(1, 2)$ and $(1', 2')$, introduced for the analyses of loading and unloading stages*

1

Previous Approaches and Motivation for the Use of the Meso-scale

In this chapter we present, in a synthetic form, some previous change of scale approaches for the definition of the mechanical behavior of granular materials. The change of scale approach based on the microscopic scale is described and limitations of this approach are put into evidence. To this end, we first present the change of scale for texture variables and then the average and localization operators for static and kinematic variables (these operators will be represented schematically in Figure 1.4).

1.1. Change of scale for texture variables

The overall behavior of a granular material is closely related to its microstructure evolution during a loading process. From this perspective, the comprehension of the fabric and its evolution is a key concept for the change of scale in granular materials. To account for the fabric in a granular sample, three types of information need to be considered: description of solid particles, description of the organization of these solid particles and description of the voids around these solid particles. The first kind of information is given by the grading curve and by some parameters characterizing the particle shape. The last type of information will be described in Chapter 3, dedicated to a description of the granular texture at the meso-scale. We present some variables for the description of the array of solid particles, usually defined by a measure of compactness and a measure of anisotropy.

1.1.1. *Coordination number and compactness*

Compactness of a granular material is defined at the REV scale by the porosity, Φ, or the void ratio, e:

$$\Phi = \frac{V_v}{V} \quad \text{and} \quad e = \frac{V_v}{V_s}, \qquad [1.1]$$

where V_v, V_s and V are the void volume, the solid volume and the total volume, respectively. At the microscopic scale the compactness can be described by the coordination number, Γ, which is the mean number of contacts for each particle in a given granular material. Several empirical relationships have been proposed in the literature, quoted in [CAM 12, CAM 10], to relate variables defined at the REV scale, (Φ, e) and variables at the microscopic scale, Γ. These relationships are only crude approximations because they do not take into account the size and shape of particles.

1.1.2. *Anisotropy*

Several tensorial quantities, called *fabric tensors*, have been proposed in the literature [SAT 82] to define the anisotropy of a set of particles in contact. These quantities are defined from local variables (Figure 1.1), as the unit normal vector, n^k, at contact point P^k, the branch vector, l^k, joining centers of mass of two particles in contact and the vector, b^g, corresponding to the maximal dimension of particule g, by the following relations:

Contact fabric tensor: $H_{ij} = \dfrac{1}{N_c} \sum\limits_{k=1}^{N_c} n_i^k n_j^k,$ [1.2]

Branch fabric tensor: $H'_{ij} = \dfrac{1}{N_c} \sum\limits_{k=1}^{N_c} l_i^k l_j^k,$ [1.3]

Combined fabric tensor: $H''_{ij} = \dfrac{1}{N_c} \sum\limits_{k=1}^{N_c} n_i^k l_j^k,$ [1.4]

Orientation fabric tensor: $H'''_{ij} = \dfrac{1}{N_g} \sum\limits_{g=1}^{N_g} b_i^g b_j^g.$ [1.5]

N_g and N_c are the number of particles and the number of contacts, respectively. Vectors n^k and l^k are defined for all contacts k, while vectors b^g are defined for all particles g. These tensors are usually defined for a set of particles in contact but tensors H and H''' can also be defined for each particle in a particle set. These tensors allow a first approximation of the orientation distributions of variables n, l and b [CAM 10]. For complex textures it can be necessary to use fourth-order tensors, H_{ijpq}, H'_{ijpq}, H''_{ijpq} and H'''_{ijpq}, to have a correct description of the different orientation distributions. The fourth-order tensor H_{ijpq} is defined as:

$$H_{ijpq} = \frac{1}{N_c} \sum_{k=1}^{N_c} n_i^k n_j^k n_p^k n_q^k.$$ [1.6]

It has been shown by [CAM 10] that the distribution of contact orientations analyzed after a shearing test (for example, the one performed on a 2D granular material by [ODA 82]) is much more accurately described by the fourth-order tensor H_{ijpq} defined in equation [1.6] than by the second-order tensor H_{ij} defined in equation [1.2].

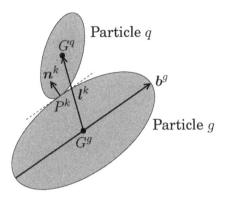

Particle q

Particle g

Figure 1.1. *Definition of the variables n^k, l^k and b^g used for the definition of anisotropy in granular materials*

1.2. Change of scale for static variables

Different approaches to the change of scale have been developed in the literature for the static variables in granular materials since the pioneering papers proposed by [LOV 44] and [WEB 66, ROT 81, CHR 81, CAM 95, BAG 96, EDW 98, CAI 02, GOL 02, FOR 03, MOR 10]. As shown in Figure 1.4, the change of scale approach considers two kinds of operators: *localization* and *average operators*. We present here some previous analyses about these two kinds of operators.

1.2.1. *Average operator*

This operator is intended for defining the stress tensor from local static variables. For the sake of brevity, we chose to present here only the approach developed by Cambou and Sidoroff based on the Virtual Work Theorem [CAM 95]. In this analysis the sample considered is supposed to be in a quasi-static state; hence, the acceleration of each particle is assumed to be negligible. The basic concepts at the macroscopic scale are those from continuum theory. Internal forces account for the use of the stress tensor, Σ, while the kinematic description involves the displacement and rotation vectors, u and ω, and the deformation tensor, E. We shall assume in the following that the stress is constant in the considered volume (REV).

At the microscopic scale, we have a collection of grains whose mass center is denoted by G^g. The strain in the material sample is defined through the individual motion of all particles, which means that for each particle g $(g = 1, ..., N)$ we have the translation, u^g, of the mass center of the particle and the rotation, ω^g, of the particle. We shall denote the contact force between two particles g and q by F^k (Figure 1.2) and we neglect the contact couples transmitted at contacts between particles.

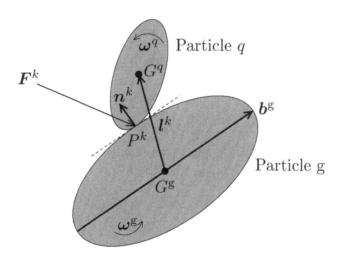

Figure 1.2. *Definition of local geometric, texture and kinematic variables used to characterize a contact between two particles whose centers of mass are denoted by G^g and G^q*

The virtual work of contact forces F^k in a virtual strain $\tilde{E}$, which gives rise to local relative displacements, $\vec{d}^k$, at contacts, can be written as:

$$\sum_k F^k \cdot \vec{d}^k, \qquad [1.7]$$

where superscript k runs over all the contacts of the sample ($k = 1, .., N_c$), F^k is the contact force acting from particle q to particle g at contact point P^k, $\vec{d}^k$ is the virtual relative displacement of particle g with respect to particle q at contact point P^k. This virtual relative displacement is defined by:

$$\vec{d}^k = \tilde{u}^g + \tilde{\omega}^g \wedge \overrightarrow{G^g P^k} - (\tilde{u}^q + \tilde{\omega}^q \wedge \overrightarrow{G^q P^k}). \qquad [1.8]$$

To relate macroscopic stress to microscopic contact forces, we use a usual identification procedure. By choosing appropriate virtual motion, we obtain:

$$\Sigma : \tilde{E} = \frac{1}{V} \sum_k F^k \cdot \vec{d}^k. \qquad [1.9]$$

To this end, we begin with a homogeneous virtual movement linear with respect to the location x:

$$\tilde{u} = \tilde{h}.x + \tilde{c}, \qquad [1.10]$$

where $\tilde{h}$ is a constant second-order tensor and $\tilde{c}$ is a constant vector. A micro-movement can be associated with equation [1.10] assuming that each particle center follows the virtual macro-movement:

$$\tilde{u}^g = \tilde{h}^S.x^g + \tilde{c}, \qquad [1.11]$$

and that the virtual rotation of each particle is equal to the virtual macro-rotation defined by $\tilde{\omega}^g \wedge y$, related to $\tilde{h}$ by

$\tilde{\omega}^g \wedge y = \tilde{h}^A \cdot y$ where y is any vector and $\tilde{h}^S$ (respectively $\tilde{h}^A$) denotes the symmetric (respectively skew-symmetric) part of $\tilde{h}$, respectively. A straightforward computation provides:

$$\tilde{d}^k = \tilde{E}.l^k \quad \text{with} \quad l^k = \overrightarrow{G^g G^{\tilde{q}}} \quad \text{and} \quad \tilde{E} = \tilde{h}^S. \qquad [1.12]$$

Combining equations [1.9] and [1.12], we get:

$$V\Sigma : \tilde{E} = \sum_k F^k \cdot \tilde{d}^k = \sum_k F^k \cdot (\tilde{E}.l^k) = \tilde{E} : \sum_k F^k \otimes l^k. \quad [1.13]$$

This last relation leads to:

$$\Sigma = \frac{1}{V} \sum_k F^k \otimes l^k. \qquad [1.14]$$

The above relation [1.14] describes the stress tensor using a sign convention used in the mechanics of solids, i.e. the stress is negative in compression and positive in extension. The symmetry of the stress tensor, Σ, defined in equation [1.14] has been analyzed by several authors [CHA 76, CAI 95, MOR 97]. They conclude that if the considered volume does not cut any particle, then the stress defined by equation [1.14] is symmetric. If some particles in the volume considered are cut by the volume boundaries, then these particles can induce a a non-symmetric part in equation [1.14]. Nevertheless, the number of such particles is usually small with respect to the total number of particles, therefore, the non-symmetric part of tensor, Σ, is in general negligible. We conclude that the stress tensor defined by equation [1.14] can usually be considered as symmetric in a first approximation. Equation [1.14] allows the stress tensor defined at the REV-level to be related to local contact forces. It is a relation which is considered to be a proven approximation of the change of scale for the static variables in granular materials considered in quasi-static states.

1.2.2. *Localization operators*

This operator is intended for defining local static variables from the macroscopic stress tensor, Σ. We present here two approaches, the first one proposed by [DEL 90, CAM 95, MAG 08] and the second one proposed by [CHA 94].

In the approach first proposed by [DEL 90], the considered static local variable is defined by:

$$f(n) = \frac{4\pi N_0 l_0}{3} P(n) \bar{F}(n), \qquad [1.15]$$

where N_0 is the number of contacts per unit volume, l_0 is the average distance between two particles in contact, $\bar{F}(n)$ is the average value of contact forces acting on a contact oriented by n and $P(n)$ is the probability of a contact oriented in direction n (Figure 1.2). Starting with the most simple situation, we can assume that there is no anisotropy due to local texture such that $f(\Sigma, n)$ is an isotropic function on Σ and n, linear with respect to Σ. In this case, the general form for $f(\Sigma, n)$ is [DEL 90, CAM 95]:

$$f(\Sigma, n) = \mu \Sigma.n + B((\Sigma.n) \cdot n)n + C \text{Trace}(\Sigma)n, \qquad [1.16]$$

where μ, B and C are constants. We have to note that for an isotropic compression stress (i.e. $\Sigma < 0$), $f(\Sigma, n)$ is oriented opposite to n and corresponds to a compression force (Figure 1.2). Using this expression in the average over the orientation, we obtain:

$$\Sigma = \frac{3}{4\pi} \int_\Omega f \otimes n \, d\Omega, \qquad [1.17]$$

where $d\Omega$ is a small solid angle centered on n and Ω is the whole solid angle corresponding to all possible orientations.

The resulting identity has to be verified identically for any Σ and this leads to the following expression for f:

$$f(\Sigma, n) = \mu\Sigma.n + \frac{1-\mu}{2}(5(\Sigma.n) \cdot n - \text{Trace}(\Sigma))n. \qquad [1.18]$$

The above expression was later extended to cover the case of an anisotropic granular sample [CAM 95] characterized by an internal traceless symmetric second-order tensor ϵ, as the use of an internal variable improves the quality of the analytical localization formula clearly. In that case, we have:

$$f(\Sigma, n) = f_0(\Sigma, n) + \frac{4}{10}(\epsilon.n - (\epsilon.n) \cdot n)n, \qquad [1.19]$$

where term $f_0(\Sigma, n)$ corresponds to the isotropic case presented in equation [1.18].

Equation [1.18], which covers the isotropic case, shows that the material parameter μ monitors the orientation of the average contact force for a fixed n. For $\mu = 0$, the contact forces are therefore oriented as n while for $\mu = 1$ the orientation is that of $\Sigma.n$. It follows that μ can be considered as directly related to the mechanical behavior of the contact, and therefore, independent of the deformation path. In particular, if the friction coefficient is close to 0 we have $\mu = 0$.

As shown by numerical DEM results [NOU 05], the shape of the distribution of average contact forces also reflects the anisotropy of particles. Later work of [MAG 08] thus proposed to introduce in the localization formula an internal variable able to account for structural anisotropy of the sample. The authors showed that the branch fabric tensor H' defined by equation [1.3] seems to be really appropriate to reflect the anisotropy of a granular sample (for example, by comparison with the use of the contact fabric tensor H defined by equation [1.2]); they obtained the following relation:

$$f(\Sigma, H', n) = \sum_{i=1}^{5} a_i g_i, \qquad [1.20]$$

with:

$$
\left\{
\begin{array}{l}
g_1 = \Sigma.n, \\
g_2 = \text{Trace}(\Sigma)n - 4(n \cdot \Sigma.n)n + 2\Sigma.n, \\
g_3 = \text{Trace}(\Sigma)(n \cdot \mathbf{dev}(H').n)n - 2(n \cdot \Sigma.n)\mathbf{dev}(H').n + ... \\
\qquad ... + \mathbf{dev}(H').(\Sigma.n), \\
g_4 = (n \cdot \Sigma.n)(n \cdot \mathbf{dev}(H').n)n - (\Sigma.(\mathbf{dev}(H').n) + ... \\
\qquad ... + \mathbf{dev}(H').(\Sigma.n))/4, \\
g_5 = \text{Trace}(\Sigma : \mathbf{dev}(H'))n + 4(n \cdot \Sigma.n)\mathbf{dev}(H') \cdot n - ... \\
\qquad -...\Sigma.(\mathbf{dev}(H').n) - 3\mathbf{dev}(H').(\Sigma.n).
\end{array}
\right.
\qquad [1.21]
$$

The theoretical tools to access the extended localization formula are the following (see [MAG 08] for more explanations):

– the principle of material frame-indifference;

– the General Representation Theorem of a vector function depending on two second-order tensors and on one first-order tensor [WAN 70];

– the linearization of function f with respect to Σ and H';

– the classical consistency condition;

– the linear independence of the functions g_i to ensure the validity of the representation.

Another approach for the static localization has been proposed by [CHA 94]. This approach is based on the following assumption:

$$
F^k(n^k) = A\hat{\Sigma} \cdot n^k, \qquad [1.22]
$$

where $\hat{\Sigma}$ denotes the modified stress tensor to account for the anisotropy of the granular sample; in particular, for an isotropic sample we have $\hat{\Sigma} = \Sigma$. Using the individual values of contact forces in the average operator defining the macroscopic stress we obtain:

$$\Sigma = \frac{1}{V}\sum_{k=1}^{N_c} \boldsymbol{F}^k \otimes \boldsymbol{l}^k = \frac{A}{V}\hat{\Sigma} \cdot \sum_{k=1}^{N_c} \boldsymbol{n}^k \otimes \boldsymbol{l}^k = A\hat{\Sigma} \cdot \boldsymbol{H}'', \quad [1.23]$$

where $\boldsymbol{H}''$ is the combined fabric tensor defined by equation [1.4] and A is a constant. In the case of spherical particles of equal diameter D, it is possible to define the constant A by considering an isotropic structure. In that case, we get [CHA 94]:

$$\boldsymbol{H}'' = \frac{D}{3}\boldsymbol{I}, \qquad A = \frac{3}{D}, \qquad \Sigma = \frac{3}{D}\hat{\Sigma} \cdot \boldsymbol{H}''. \qquad [1.24]$$

It follows that, if $\boldsymbol{G}''$ denotes the inverse of $\boldsymbol{H}''$, the localization operator can be expressed as:

$$\boldsymbol{F}^k(\Sigma, \boldsymbol{n}^k) = \Sigma.(\boldsymbol{G}''.\boldsymbol{n}^k). \qquad [1.25]$$

1.3. Change of scale for kinematic variables

For kinematic variables, as for static ones, localization and average operators have to be considered. We present here essentially previous works on average operators. The local kinematic variables usually considered are the displacement of the mass center, G^g of particle g, denoted by u^g and the rotation of particle g, denoted by ω^g (Figure 1.2). Contact between two particles g and q can be characterized either by the double subscripts, (g, q), or by subscript k which will be used to simplify the notations. The unit normal vector at a given contact is oriented from g to q (Figure 1.2). For particles in contact, considered as rigid, the relative displacement at contact P^k between particles q and g can be defined by:

$$\delta c_i^k = (\delta u_i^q - \delta u_i^g) + \epsilon_{ijl}(\delta\omega_j^q R_l^q - \delta\omega_j^g R_l^g), \qquad [1.26]$$

where δc_i^k is the relative displacement between contact point, P^{kq} belonging to particle q and contact point P^{kg} belonging to particle g, ϵ_{ijl} is the permutation symbol, R_l^g is the vector joining G^g to P^k, and R_l^q is the vector joining G^q to P^k. The local phenomena occurring at the local-scale during the deformation of granular material are complex. Different approaches have been proposed in the literature to establish the link between particle-level displacements and macro-level deformations. We will present here two approaches, the first one based on the definition of an equivalent continuum, and the second one based on the definition of the best fit approximation of the local discrete displacements.

1.3.1. *Average operator: definition of strain from a local equivalent continuum*

Different approaches have been proposed by different authors, by [BAG 93, BAG 96, KRU 96, KRU 14] in particular. In this book, we chose to only present the method proposed by [DED 00, CAM 00]. This formulation is valid for 2D analysis of particles with an arbitrary convex shape. The first step is to consider all the branch vectors joining particles in contact. These segments allow adjacent polygons to be defined; these polygons cover the total area of the considered sample. Each polygon can then be divided into several triangles. This discretization is called the Delaunay triangulation (Figure 1.3).

This decomposition is not unique, but triangles with side dimensions of the same order of magnitude have to be preferred. Each kind of discretization leads to very similar results (equal in a first approximation). Each triangle has three sides, each of which joins either two particles in contact or two neighboring particles without contact. Each branch with or without contact is denoted by l'^k. It is assumed that the strain increment, denoted by $\delta \varepsilon^e$, is constant for each triangle e.

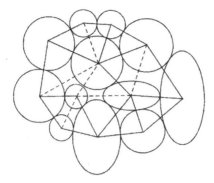

Figure 1.3. *Definition of local polygons from branches linking two particles in contact (straight lines). These polygons are divided into triangular elements, e, using dashed lines joining neighboring particles without contact. All segments in this figure (straight and dashed lines) are denoted by l'^k [CAM 10]*

As the strain is constant in each element, the displacement $u(x)$ is an affine function with respect to spatial variable x:

$$\delta u_i^e = \delta a_{ij}^e x_j + \delta u_{0i}^e, \qquad [1.27]$$

where δa_{ij}^e and δu_{0i}^e are the displacement gradient tensor and a constant vector, respectively. In each triangular element, each of the vertices can be defined by numbers 0, 1 or 2. Vertex 0 is randomly chosen as a reference. The relative displacement between two vertices $0-1$ or $0-2$ can be computed as:

$$\delta u_i^{v0} = \delta u_i^v - \delta u_i^0 = \delta a_{ij}^e (x_j^v - x_j^0) = \delta a_{ij}^e x_j^{v0}, \qquad [1.28]$$

where v is the vertex number which can take the value either 1 or 2, and $x^{v0} = x^v - x^0$ is the vector joining vertex v to the reference vertex 0. We define two vectors y^{v0} for $v = 1$, 2 such that:

$$\sum_v x_i^{v0} y_j^{v0} = \delta_{ij}, \qquad [1.29]$$

where δ_{ij} is the Kronecker symbol. By solving the system of four linear independent equations in [1.28], the displacement gradient tensor, δa^e, can be obtained as:

$$\delta a_{ij}^e = \sum_v \delta u_i^{v0} y_j^{v0}. \qquad [1.30]$$

It should be noted that the displacement gradient tensor, δa^e, obtained in equation [1.30] does not depend on the chosen reference vertex because in each element the local displacement field, $u(x)$, is an affine function with respect to the spatial variable, x.

The strain tensor, $\delta \varepsilon^e$, can be obtained from the displacement gradient tensor, δa^e:

$$\delta \varepsilon_{ij}^e = \frac{1}{2}(\delta a_{ij}^e + \delta a_{ji}^e). \qquad [1.31]$$

Relations [1.30] and [1.31] put clearly in evidence the dependency of the local strain with respect to the local texture characterized by vectors y^{v0}.

The strain increment defined at the sample scale, δE, is then obtained as the volumetric average value of $\delta \varepsilon^e$ over all triangles in the Delaunay discretization:

$$\delta E_{ij} = \frac{1}{V} \sum_{e=1}^{N_e} \delta \varepsilon_{ij}^e V^e, \qquad [1.32]$$

where N_e is the number of triangles and V^e is the area of triangular element e.

1.3.2. *Average operator: strain defined from best-fit methods*

In these methods the increment of displacement gradient tensor, δA, defined in the REV is optimized in such a way that the theoretical displacement field linked to δA is the best-fit to some actual local displacement field.

The main difference between all the best-fit methods lies in the considered local displacement field [CAM 00, DED 00]. In the method proposed by [CUN 79], the considered local displacement field is the displacement of the centers of neighboring particles. In the method proposed by [LIA 97] the local displacement field is the relative displacement at contacts between particles. In the method proposed by [CAM 00], the local displacement field is the relative displacement of the centers of neighboring particles. In this book, we chose to present this last method only. As for the other formulations, a displacement gradient tensor, δA, defined at the REV level is considered. A Delaunay triangulation based on the centers of particles allows local triangular elements between centers of neighboring particles to be defined. Two particles are considered to be neighbors if their centers are connected by a branch and they are not necessarily in contact. In the equivalent continuum, the relative displacement between the two centers of neighboring particles can be estimated by $\delta A_{ij} l'^{k}_{j}$, with l'^{k}_{j} being the branch vector linking two centers of the considered particles. For each pair of neighboring particles we can compare the actual relative displacement of the two particle centers, $\delta l'^{k}_{i}$ and the theoretical one, $\delta A_{ij} l'^{k}_{j}$, derived from the equivalent continuum. The deviation between these two fields can be computed as:

$$Z = \sum_{k=1}^{N'_c} (\delta A_{ij} l'^{k}_{j} - \delta l'^{k}_{i})(\delta A_{ij} l'^{k}_{j} - \delta l'^{k}_{i}), \qquad [1.33]$$

where N'_c is the number of pairs of neighboring particles defined by the Delaunay triangulation. Minimization of Z with respect to the tensor δA leads to the following equation:

$$\frac{\partial Z}{\partial (\delta A_{pr})} = \sum_{k=1}^{N'_c} (\delta A_{pj} l'^{k}_{j} - \delta l'^{k}_{p}) l'^{k}_{r} = 0. \qquad [1.34]$$

This equation leads to:

$$\delta A_{pj} \sum_{k=1}^{N_c'} l'^k_j l'^k_r = \sum_{k=1}^{N_c'} (\delta l'^k_p) l'^k_r. \qquad [1.35]$$

Let tensor G be defined by:

$$G_{ij} \hat{H}_{jr} = \delta_{ir} \quad \text{with} \quad \hat{H}_{ij} = \frac{1}{N_c'} \sum_{k=1}^{N_c'} l'^k_i l'^k_j. \qquad [1.36]$$

The increment of the displacement gradient can then be defined by:

$$\delta A_{ij} = \frac{G_{ir}}{N_c'} \sum_{k=1}^{N_c'} (\delta l'^k_r) l'^k_j. \qquad [1.37]$$

The strain increment can be then defined as the symmetric part of the increment of the displacement gradient tensor given by equation [1.37].

[BAG 06] and [CAM 00] have analyzed, on the basis of numerical DEM simulations, different approaches used in the literature to compute strains at the REV scale from local kinematic variables. The conclusions of these analyses are the following:

– all the methods considering an equivalent continuum strain as those proposed by [BAG 93, BAG 96, KRU 96, CAM 00] are in good agreement with the strain measured at the macroscopic level (sample scale);

– among the different best-fit methods, only the methods proposed by [CUN 79] or by [CAM 00], which consider the relative displacements of the centers of neighboring particles, give a correct estimate of the strain measured at the sample scale;

– the best-fit method proposed by [LIA 97], which considers the relative displacements at each contact, underestimates the strain at the sample scale.

It is clear that in a granular material, local relative displacements between particles are very different if we consider contacting particles or neighboring particles without contact. In the former, compressive normal displacements are not possible (or are very small, linked to local contact deformations) and in the latter, these kinds of displacements are possible. For this reason, methods considering only contacting particles are not able to give a correct estimate of strain at the sample scale, unlike methods considering also the neighboring particles.

1.3.3. *Localization operator: definition of local kinematic variables from the strain tensor defined in a REV*

Different authors [CHA 94, CAM 95] have considered the increments of relative displacements, δc_i^k, of two particles at their contact point as the local kinematic variables. These local variables are defined from the increment of the macro-strain tensor, δE_{ij}, using the following localization operator:

$$\delta c_i^k = \delta E_{ij} l_j^k, \qquad\qquad [1.38]$$

where l_j^k denotes the branch vector joining the two centers of particles in contact (contact k). We have seen in the previous sections that the strain defined at the REV is not only related to the relative displacements of particles in contact but also to the relative displacements of neighboring particles. Therefore, equation [1.38] is not correct to define the local kinematic variables. [CAM 95] have proposed more complex formulations for this operator. These formulations take into account internal parameters which are complex to define. At this time, the definition of a relevant localization operator for the kinematic variables remains an open issue.

1.4. Statistical homogenization in granular materials

The main objective of the homogenization method is to obtain a constitutive relation at the REV scale from information on the material behavior at the micro-scale and from the microstructure. For granular materials, the micro-scale is generally the grain scale. The REV scale is several orders of magnitude higher. The homogenization methods can be used either for periodic media or for arbitrary disordered media; the latter case can be called statistical homogenization. The homogenization process is based on three relations, depicted in Figure 1.4:

– a localization operator;

– a local constitutive law;

– an average operator.

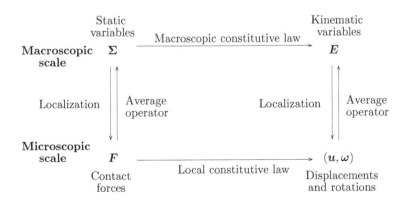

Figure 1.4. *General homogenization operators considering a micro-scale for granular materials*

An intrinsic difficulty in the case of granular materials comes from the fact that the variables at the micro-scale and at the macro-scale are of a different nature. At the macro-scale, the behavior law uses tensorial variables (stress tensor, strain tensor), while at the micro-scale the material behavior is derived using vectorial variables (contact forces, grain displacements, grain rotations). The average operators have been defined in several papers [WEB 66, CHA 76, CHR 81, BAG 93, CAM 95, CAI 95, LIA 97, CAM 00, BAG 06, MOR 10] and the main difficulty remains the definition of a localization operator. Available results deal mainly with the elastic behavior of granular samples.

To illustrate this problem we present the approach proposed by [CHA 94], designed to model the elastic behavior of granular samples. This method is based on a kinematic localization. According to Figure 1.4, the kinematic localization operator defined by equation [1.38] is used to obtain the field of local relative displacements at contact points. Then, the local contact law gives the value of contact forces. Finally, the average over the sample provides the value of the stress tensor. The authors assume local elastic behavior at contact k in the simple form:

$$\delta F_i^k = K_{ij}^k \delta c_j^k, \qquad\qquad [1.39]$$

where tensor K_{ij}^k represents local elasticity of the contact k. To obtain the macro-stress in volume V, the usual averaging operator [1.14] defined by [WEB 66, CHA 76, CHR 81, BAG 93, CAM 95, CAI 95, LIA 97, MOR 10] is used:

$$\delta \Sigma_{ij} = \frac{1}{V} \sum_k \delta F_i^k l_j^k. \qquad\qquad [1.40]$$

Then relations [1.38], [1.39] and [1.40] together yield to the incremental constitutive law:

$$\delta \Sigma_{ij} = \frac{1}{V} \delta E_{mp} \sum_k \delta K_{im}^k l_p^k l_j^k, \qquad [1.41]$$

which defines the fourth-order Hooke's tensor as:

$$C_{impj} = \frac{1}{V} \sum_k K_{im}^k l_p^k l_j^k. \qquad [1.42]$$

[CHA 94] gave the expression of this fourth-order tensor in the case of a granular sample containing spherical particles of equal diameter, in an isotropic state. The overall behavior is then defined using a Young's modulus, E and a Poisson's ratio, ν:

$$E = \frac{N_0 D^2}{3} k_n \frac{2 + 3\alpha}{4 + \alpha} \quad \text{and} \quad \nu = \frac{1 - \alpha}{4 + \alpha}, \qquad [1.43]$$

where $\alpha = k_t / k_n$ is the ratio of the tangential stiffness, k_t, to the normal stiffness, k_n, D denotes the mean particle diameter and N_0 is the number of particles in a unit volume.

As demonstrated in [CAM 00] and [BAG 06], it can be noted that equation [1.38] linking local relative displacements to the global strain defined at the REV level is, in general, not acceptable, so this approach leads to unrealistic results. In the same paper [CHA 94], the authors proposed another estimate of the elasticity of a granular sample based on static localization. A static localization operator provides the contact forces in terms of macroscopic stress and the local contact law gives the relative displacements. Finally, an average operation gives the macroscopic strain from the local relative displacements. Results obtained from this method have the same drawbacks as the previous method because in this model the global strain at the REV level is computed from the relative displacements at contacts between particles.

Other authors [CAM 95, EME 96] have proposed similar models but with some different localization operators. Nevertheless, these models present the same drawbacks as the model presented above.

1.5. Difficulties and limitations for statistical homogenization in granular materials using a micro-scale

As we have seen at the beginning of the last section, a statistical homogenization approach is built using three operators: a localization operator, a local behavior law and an average operator. In this type of material behavior description, the local information is not explicit and only statistical data are available. It follows that the description is only approximate and so will be the corresponding solution. Two variants are possible: (1) assuming a localization operator for the kinematic variables and an average operator for the static variables and (2) assuming a localization operator for the static variables and an average operator for the kinematic variables. The solutions obtained using the aforementioned procedures are, in general, not the same. In some special cases, for example in the case of a linear elastic local behavior, it has been demonstrated that the mentioned procedures lead to upper and lower bounds for the macroscopic response. In the general case, the obtained solution is only a rough approximation. The analysis developed in several papers [CAM 00, BAG 06] leads to the following conclusions:

– the static averaging relation defined by equation [1.14] is well-established and recognized;

– there is no clear consensus concerning the static localization operator. The relations proposed in [CAM 95] were found to be in good agreement with results given by numerical simulations;

– several kinematic average equations were proposed in the literature. The key point is the choice of the local kinematic variables and the formulation used. Numerical simulations seems to favor equation [1.37], which is both simple and realistic. We note that this formulation uses the relative displacement of neighboring particles at the local level. Other relations based on an equivalent continuum can also be used [BAG 96, KRU 96]. The kinematic localization relation is, in some common cases, written in the form of equation [1.38]. The results obtained from numerical simulations show that this relation does not allow the actual behavior to be approached in a realistic manner;

– definition of a relevant localization operator for kinematic variables remains an open issue.

Behind the inherent difficulties concerning the localization and average operators, the statistical homogenization procedure needs a local behavior law and a choice of local variables. This is not always an easy task. Most of the contact models relate the contact force to the relative displacement. These models particularly impose the choice of the local kinematic variable to be the relative displacement at contacts. As mentioned, this choice is not a correct one when considering the kinematic average, leading to unrealistic results. We may, therefore, conclude that the statistical homogenization based on a micro-scale defined at the level of the contacts between particles does not provide realistic results.

Therefore, if we want to use a change of scale approach to define the behavior of granular material it is necessary to propose another approach. Several authors have proposed the use of a meso-scale based on local arrays. [LIA 97] consider local groups of particles composed of a central particle and of particles in contact with this particle. [SAB 97] uses a partition of a granular sample into local tetrahedra which are considered as the meso-domains. [NIC 11] consider hexagonal arrays of particles. In these models, the local constitutive law is defined in these meso-domains. The approaches mentioned present drawbacks but are interesting; we will, thus, develop in this book a new approach based on the same initial point of view, that is to say the definition of local meso-domains in which a local constitutive law will be defined.

2

Definition of a Meso-scale for Granular Materials

2.1. Brief presentation of the Discrete Element Method

Analyses presented in this book require local information at the particle level. The Discrete Element Method (DEM) appears to be suited to this kind of data. The DEM has been widely used to model granular media since the pioneering work of [CUN 79, MOR 93, THO 00, LUD 08, RAD 09, OSU 11]. It consists of modeling a granular assembly as a collection of rigid distinct particles which interact with each other through contacts. The motion of each particle is described by the Newton-Euler dynamic equations of motion and the interaction between particles is described by a contact model. Modeling interaction between particles is a hard task, which differs between the molecular dynamics (MD) approach and the contact dynamics (CD) approach.

The MD approach assumes that two particles in contact can overlap slightly, and that hence, the contact forces can be determined uniquely from the contact relative displacements [CUN 79, THO 00, LUD 08]. Several force-displacement relations have been proposed such as the linear elastic model and the Hertz model. According to the first one, the normal and tangential contact forces, f_n^k and f_s^k, are linearly related to the normal and tangential relative displacements, c_n^k and c_s^k, by a normal and a tangential stiffness, k_n and k_s, respectively (Figure 2.1). The tangential interaction must respect Coulomb's Law of Friction $|f_s^k| \leq \mu f_n^k$ with friction coefficient μ.

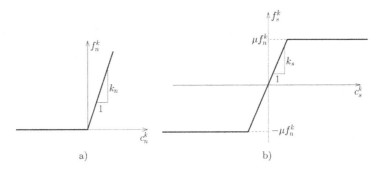

Figure 2.1. *Linear elastic model used in the MD approach: a) for the normal contact force and b) for the tangential contact force*

In the CD approach, particles are assumed to be perfectly rigid so the compliance at the contact between two particles is neglected [MOR 93, JEA 99, RAD 09]. As a result, the mono-valued graphs shown in Figure 2.1 become the multi-valued graphs displayed in Figure 2.2. Neglecting the particle compliance leads to a discontinuity in particle velocities when two particles collide. Therefore, Newton's restitution law with normal and tangential restitution coefficients, e_n and e_s, is used to calculate the jump in relative velocities:

$$\dot{c}_n^{k+} = -e_n \dot{c}_n^{k-}, \quad \dot{c}_s^{k+} = -e_s \dot{c}_s^{k-}, \tag{2.1}$$

where the signs $+$ and $-$ represent the post- and pre-impact velocities, respectively.

The system of particle motion equations and contact force-displacement relations are integrated numerically in a time-stepping manner. In the MD approach, state variables (the particle position and velocity) are smooth, and therefore, an explicit high-order integration scheme such as the central difference scheme [OSU 11] can be used. However, a sufficiently small time step Δt is required to guarantee the integration stability. In the framework of the MD method, a granular assembly can be thought of as being a system of masses and springs. For such a system, the state variables oscillate with different eigenfrequencies. The numerical stability requires that the time step Δt must be much smaller than the smallest oscillation period, and hence, is proportional to $\sqrt{m/k_n}$ with particle mass m. It follows that the time step Δt decreases as the particle stiffness k_n increases and the particle mass m decreases. To model a granular material, the particle stiffness is, in general, high and the particle mass is small, leading to a very small time step Δt. In the CD approach the state variables are non-smooth and the contact forces can be implicitly determined using an iterative procedure only. As a result, the integration scheme is implicit and of low-order [JEA 99, RAD 09]. Nevertheless, the time step Δt in the CD approach is, in general, considerably bigger than in the MD approach. The jump in velocity when two particles collide leads to a great fluctuation of local variables such as the positions of particle centers; hence some averaging operations are often necessary to obtain a measure of these variables in the CD approach. In both approaches, some damping parameters are used to speed up the convergence of the system to the equilibrium state.

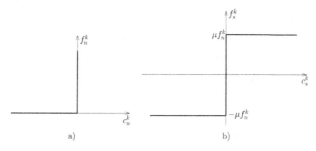

Figure 2.2. *Contact model used in the CD approach: a) for the normal contact force and b) for the tangential contact force*

2.2. Description of the four numerical granular samples

For the purpose of our study, four different 2D granular samples are simulated using the DEM. The first and second samples, A and B, are simulated with the MD approach using the software PFC [ITA 99]. These two samples are both composed of circular particles but the normal and tangential contact stiffnesses, k_n and k_s, used for sample B are much higher than the ones used for sample A. A loading/unloading biaxial test is performed on sample B, while only a loading biaxial test is performed on sample A. The two other samples, C and D, are simulated with the CD approach using the open source platform LMGC90 [DUB 06]. Sample C is composed of circular particles, while sample D is composed of octagonal particles. Biaxial loading tests are performed on these samples. The four samples are generated without considering gravity forces. Moreover, during the generation and the compaction stages of these samples, the interparticle friction angle and the friction between particle and walls are set to zero. Then, during the biaxial tests, the friction between particles and walls remains equal to zero, while the interparticle friction angles are set at non-zero values, specific for each sample.

2.2.1. Description of sample A

Sample A is composed of 25,000 circular particles with a uniform size distribution from 4–8 mm before compaction. Particles are inserted in a random way into a rectangular box composed of four rigid frictionless walls (Figure 2.3). The normal and tangential stiffnesses, k_n and k_s, are equal to 5.0×10^{-7} N/m. The contact friction angle, ψ, is equal to 30° (friction coefficient, $\mu = \tan \psi$). At this stage, the sample is very loose. It is then compacted by gradually expanding each particle. By doing so, the stresses Σ_{11} and Σ_{22} in the sample increase and the expansion ends when both stresses are 100 kPa. At the end of the compaction, the minimum and maximum particle diameters are 4.8 mm and 9.6 mm, respectively.

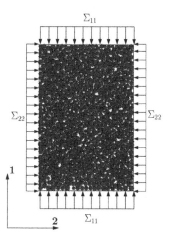

Figure 2.3. *Illustration of a biaxial test performed on a numerical sample. The longitudinal and transverse directions are denoted by* 1 *and* 2*, respectively*

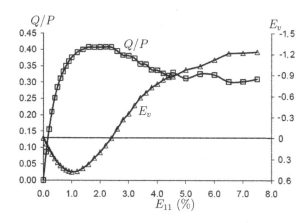

Figure 2.4. *The stress ratio,* Q/P, *and the volumetric strain,* E_v, *of sample A versus the axial strain,* E_{11}, *during the biaxial compression test*

The sample after compaction is quite dense with a porosity Φ of 0.158. A biaxial compression is then applied to the sample by prescribing a compression strain rate $\dot{E}_{11} = 0.01$ s^{-1} in the longitudinal direction while keeping the lateral stress, Σ_{22}, constant. Figure 2.4 shows the stress ratio Q/P and the volumetric strain, E_v, of the sample during the biaxial compression test. In this figure, Q and P are the deviatoric stress and the mean stress in the sample, respectively:

$$Q = \frac{\Sigma_1 - \Sigma_2}{2} \text{ and } P = \frac{\Sigma_1 + \Sigma_2}{2},$$ [2.2]

and E_v is the macroscopic volumetric strain:

$$E_v = E_1 + E_2.$$ [2.3]

It should be noted that for biaxial loading/unloading tests, directions 1 and 2 are the major or the minor principal directions of the stress and strain tensors; hence $\Sigma_1 = \Sigma_{11}$, $\Sigma_2 = \Sigma_{22}$, $E_1 = E_{11}$ and $E_2 = E_{22}$. As shown in Figure 2.4, sample A presents the typical behavior of dense granular samples as observed experimentally. The characteristic state (at which the sample changes from contraction to dilation), the peak state (at which the stress ratio Q/P is maximum) and the critical state (at which the stress ratio Q/P remains constant) occur at $E_{11} = 1\%$, 2% and 7%, respectively. It is worth noting that this sample contracts significantly up to $E_{11} = 1\%$ despite its high initial density. This is due to the fact that the value $k_n = 5.0 \times 10^{-7}$ N/m chosen in this simulation is relatively low ($k_n/\Sigma_{22} = 500$), leading to an important overlap at contacts between particles and then a high elastic deformation of the sample. In the numerical simulations performed by [ROU 02], $k_n/\Sigma_{22} = 10^5$ was chosen to consider particles as being nearly rigid. However, simulation with such a high value of k_n is very time-consuming.

Figure 2.5 shows two polar representations of the contact orientation distributions at the initial state and at the peak state for sample A. At the initial state, the sample is isotropic in view of the compaction method by diameter expansion used in the simulation. It is well-known that when a granular assembly is subjected to a deviatoric loading, contacts between particles tend to be oriented in the major principal stress direction, causing the anisotropy of the sample. As shown in Figure 2.5, contacts in sample A are preferentially oriented in the compression direction at the peak state.

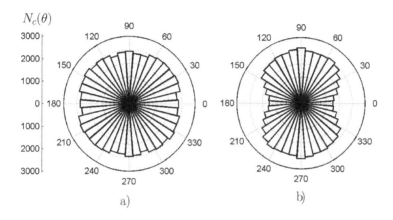

Figure 2.5. *Orientation distribution of contact normals a) at the initial state and b) at the peak state for sample A*

The fabric tensor, H, defined by equation [1.2] and the coordination number, Γ, are used to study the evolution of internal structure of sample A during the compression test. During the simulated biaxial compression test, the fabric tensor, H, is almost diagonal and the component H_{11} increases whereas the component H_{22} decreases. Consequently, the anisotropy of the sample can be described by the ratio H_{11}/H_{22}. Figure 2.6 indicates the evolution of the ratio H_{11}/H_{22} and of the coordination number Γ during the simulated test. The increase in the ratio H_{11}/H_{22} means that the anisotropy of the sample increases during the deviatoric loading. We can also remark that the sample reaches the maximum anisotropy after reaching the maximum stress ratio Q/P. The coordination number, Γ, almost decreases during the test, which is essentially due to the dilatancy. In the initial stage, the contractancy leads to a very small increase of Γ because this contractancy is essentially due to a high elastic deformation at contact points.

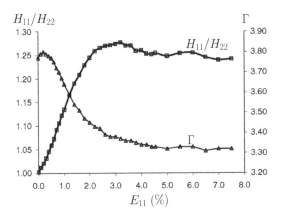

Figure 2.6. *The fabric ratio, H_{11}/H_{22}, and the coordination number, Γ, versus the axial strain, E_{11}, for sample A*

2.2.2. *Description of sample B*

Sample B is identical to sample A except that the contact stiffnesses, k_n and k_s, are equal to 10^9 N/m. This high value of k_n and k_s significantly reduces the overlap between particles which allows a focus on plastic deformations; however, the resulting step size Δt is very small and the computation time is very long. This sample is generated and compacted by the same method as the one used for sample A. After compaction, it is slightly looser than sample A (the porosity is about 0.167 for sample B as compared to 0.158 for sample A). A biaxial compression is first applied on this sample until an axial deformation $E_{11} = 10$ %, followed by a biaxial extension test, here called the *unloading*, during which the axial stress Σ_{11} is progressively reduced while the lateral stress Σ_{22} is kept constant.

Figure 2.7 shows the stress ratio Q/P and the volumetric strain E_v of sample B during the performed loading/ unloading test. It can be clearly seen that the stress ratio Q/P at the peak state for this sample is lower than for sample A, due to its lower density at the initial state. The high value of the particle stiffnesses reduces the elastic deformation of this sample effectively, the elastic contractancy occurring at the initial stage of loading in particular. The sample contracts strongly at the initial stage of the unloading and then dilates as the unloading continues. Figure 2.8 shows the anisotropy induced by the loading and unloading. During the unloading, the direction of the anisotropy changes gradually from the longitudinal direction to the transverse one, which becomes the major principal stress direction at the end of the unloading stage.

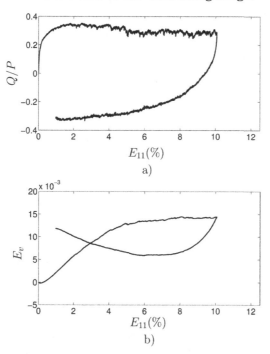

Figure 2.7. *Macroscopic behavior of sample B during the loading/unloading test: a) the stress ratio Q/P and b) the volumetric strain E_v versus the axial strain, E_{11}*

2.2.3. Description of samples C and D

Sample C contains 5,000 poly-sized circular particles and sample D contains 5,000 poly-sized octagonal particles. They are simulated with the CD approach. Inelastic collisions between particles are assumed, which indicates that the normal and tangential restitution coefficients, e_n and e_s, are equal to zero. Both samples have the same uniform size distribution with particle diameter varying from 6–19.6 cm. Octagonal particles are defined in Figure 2.9. As shown in this figure, two circles are considered with radii R and $R - \alpha R$, respectively. Each octagonal particle is symmetrical with respect to diameter AE. Lines AH, AB, ED and EF are tangent to the inner circle. In this study, α is equal to 0.2.

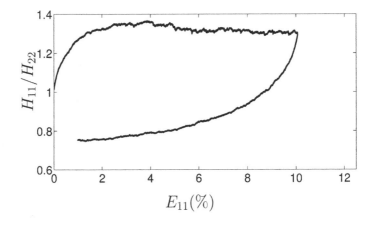

Figure 2.8. *The fabric ratio, H_{11}/H_{22}, versus the axial strain, E_{11}, during the loading/unloading test for sample B*

A loose packing composed of disks is generated first. Small constant velocities are then applied on the upper and right boundaries to compact the samples. During this stage, the friction between particles and between particles and walls is set to zero. This compaction phase is applied until an isotropic stress equal to 10 kPa is reached. The assembly of octagonal particles is then obtained from the compacted sample of circular particles, by replacing each circular particle with an octagonal particle with the same radius R. The orientation of each octagonal particle is chosen randomly. As the obtained sample is not dense enough due to the particle shape, the same compaction procedure as previously defined is applied until obtaining an isotropic stress equal to 10 kPa. Figure 2.10 shows a zoom inside sample D after compaction. After the compaction, the porosity, ϕ, is equal to 0.158, for sample C and to 0.127 for sample D. A biaxial compression test is then simulated on these two samples by applying a constant stress $\Sigma_{22} = 10$ kPa on the two lateral walls and a constant velocity on the upper wall. During this stage, the interparticle friction angle is set to $27°$.

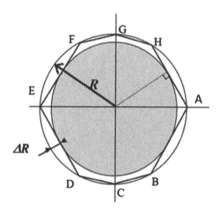

Figure 2.9. *Geometrical definition of octagonal particles for sample D*

Figure 2.11 shows the stress ratio, Q/P, and the void ratio, e, versus the axial strain E_{11} for the two biaxial compression tests simulated on samples C and D. As these simulations are based on the CD approach, collisions between rigid particles lead to fluctuations of contact forces and of local velocities. These quantities are, therefore, averaged over 500 time steps before computing global stress and strain tensors. A high rigidity of these samples at the beginning of loading and a lack of initial contractancy can be observed. These phenomena are due to, firstly, rigid particles assumed in the CD approach, and secondly, the high density of samples C and D. It is interesting to note that with the same value of local contact friction ($\psi = 27°$), the same grading curve, and the same process of generation, sample D composed of octagonal particles presents considerably greater internal friction angles at the peak stress and at the critical state ($24°$ and $20°$, respectively) than sample C composed of circular particles ($20°$ and $15.3°$, respectively). In addition, sample D shows a greater dilatancy than sample C, which explains the higher internal friction angle of sample D. These results are in good agreement with experimental studies stating that particle shape has a great influence on the mechanical behavior of granular materials [GUO 07].

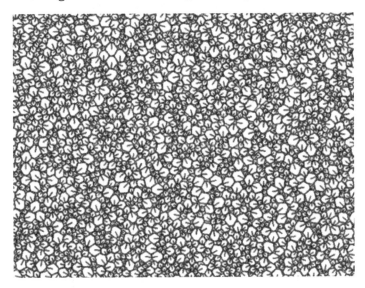

Figure 2.10. *A zoomed-in view of sample D composed of octagonal particles*

Most of the following results are obtained with sample A. If a result is presented without any specification, it means that it is obtained with sample A. The three other samples will be explicitly specified when they are considered. Sample B is essentially used to analyze the local behavior during the unloading stage and the plastic deformation since the elastic deformation is very small for this sample. The effects of the particle shape and of the modeling method are analyzed with samples C and D.

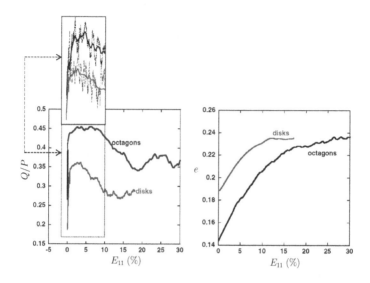

Figure 2.11. *a) The stress ratio, Q/P, and b) the void ratio, e, versus the axial strain, E_{11}, for samples C and D*

2.3. Meso-scale for 2D granular media

Several geometrical representations of a 2D granular assembly have been proposed in the literature. A *Delaunay network* (Figure 2.12) consists of connecting the centers of every three neighboring particles [BAG 96, CAM 00, SAT 04]. In such a system, a *closed branch* (represented by the solid line) connects two particles in contact, whereas an *open branch* (represented by the dashed line) connects two particles without contact. The dual graph of a Delaunay network is called the *Dirichlet system* (Figure 2.12) which entirely sub-divides a granular medium into polygonal cells called *Dirichlet polygons*. For a given particle, the Dirichlet polygon associated with it is the set of points having a shorter or equal tangent to it than to any other particles.

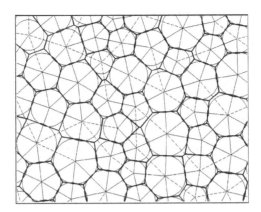

Figure 2.12. *Delaunay network and Dirichlet system constructed for an assembly of circular particles. Dashed and solid lines depict branches joining two particles without and with contact, respectively*

[SAT 04] proposed a method to construct a Dirichlet system from a Delaunay network. For this method, a Dirichlet center D is first defined for each triangle obtained in the Delaunay triangulation as follows (Figure 2.13(a)):

$$x_D = \frac{1}{2V^e} \sum_{i=1}^{3} c^i b^i, \qquad [2.4]$$

where:

– V^e is the area of triangle e;

– scalar c^i is defined for each particle i: $c^i = (r^{i^2} - x^i . x^i)/2$ with r^i and x^i being its radius and the position of its center;

– index i is also given to the side of the triangle opposite to particle i;

– vector b^i is outward normal to edge i and its magnitude is equal to the length of edge i.

The Dirichlet cell associated with a particle is the polygon whose vertices are Dirichlet centers, defined for all the triangles taking the center of the considered particle as a vertex (Figure 2.13(b)).

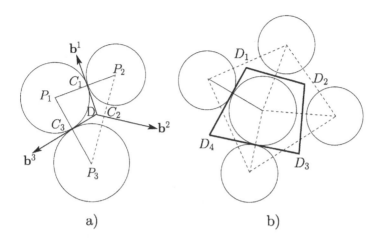

a)   b)

Figure 2.13. *a) Dirichlet center defined for a triangle and*
b) Dirichlet cell associated to a particle

The use of Dirichlet cells allows a complete discretization of a 2D granular material. Another approach can be considered to discretize a 2D granular material. Starting from the Delaunay network constructed for a granular assembly (Figure 2.14(a)), we associate together every two neighboring triangles which share an open branch. By doing so, the granular assembly can be entirely subdivided into *meso-domains* (otherwise called *loops*), each of which is delimited by closed branches as shown in Figure 2.14(b). This system is called the *particle graph* [SAT 78, SAT 92, KUH 97]. Note that the particle graph can be constructed for any particle shape. Figure 2.15 shows part of the particle graph constructed for an assembly of polygonal particles [CHA 14].

During our numerical simulations, gravity is not applied to each particle. As a consequence, a significant number of particles in the assembly has either contact or only one or two contacts. These particles are called *rattlers* and they do not participate in supporting the external loading. When constructing the particle graph for an assembly at a given value of the axial strain, E_{11}, all the rattlers are temporarily removed from the assembly [KUH 99]. During the loading, a rattler can be subsequently activated and will then be included in the particle graph.

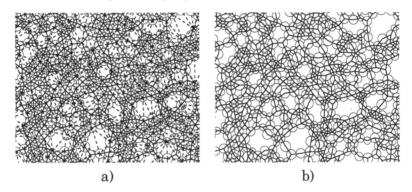

a)   b)

Figure 2.14. *a) Associating triangles in Delaunay network to obtain b) the particle graph for a loose granular sample composed of disks [NGU 09a]*

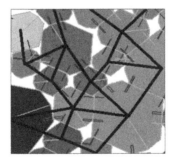

Figure 2.15. *Part of the particle graph constructed for an assembly of polygonal particles of sample D [CHA 14]*

Particle positions and contacts change during the loading. Therefore, the particle graph must be updated throughout the loading. In the current study, this update is performed for every 0.1% of the axial strain E_{11}.

Meso-domains are polygons with different numbers of edges, as illustrated in Figure 2.16. Let the valence, r^m, of a meso-domain be its number of edges. The valence, r^m, is then equal to 3 for triangles, 4 for quadrilaterals, 5 for pentagons, etc. The composition of a particle graph is represented by the volume fraction occupied by each set of meso-domains having the same valence (set of triangles, set of quadrilaterals, set of pentagons, etc.). Figure 2.17 shows the composition of the particle graph constructed for sample A at different loading levels. At the initial state, meso-domains of low valence (3, 4 and 5) are the most dominant in the particle graph. During the biaxial compression, some meso-domains of low valence are destroyed and new meso-domains of high valence (6, 7, 8, etc.) are created. Furthermore, the composition of the particle graph seems to evolve slowly after the peak state at the axial strain $E_{11} = 2\%$. A similar composition is observed for the particle graphs constructed for samples C and D (these samples are simulated with the CD approach) as indicated in Figure 2.18. One can also remark that the evolution of the particle graph constructed for sample D composed of octagonal particles is more marked than that for sample C composed of circular particles.

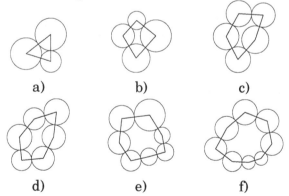

a)   b)   c)

d)   e)   f)

Figure 2.16. *a) Triangle, b) quadrilateral, c) pentagon, d) hexagon, e) heptagon and f) octagon in the particle graph constructed for sample A*

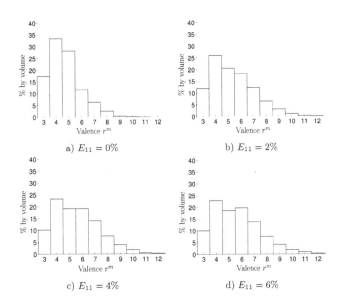

Figure 2.17. *Composition of particle graph constructed*
for sample A at different loading levels

The redundancy number, $\hat{r}^m$, of a meso-domain is defined as
$\hat{r}^m = r^m - 3$. The mean redundancy number $\hat{R}$ of an assembly
is:

$$\hat{R} = \frac{1}{N_m} \sum_{m=1}^{N_m} \hat{r}^m, \qquad [2.5]$$

with N_m number of meso-domains. If the particle graph is
composed of only triangles the mean redundancy number, $\hat{R}$,
is zero. As shown in Figure 2.19, the mean redundancy
number $\hat{R}$ increases with the axial strain, E_{11}, especially at
the onset of the loading. This confirms the loss of low-valence
meso-domains and the creation of high-valence meso-domains
during the deviatoric loading. Let Γ^* be the modified
coordination number of a sample when all rattlers are
removed. The evolution of Γ^* is plotted in Figure 2.19. We can

observe a strong correlation between $\hat{R}$ and Γ^*: Γ^* decreases, while $\hat{R}$ increases. This correlation indicates that the loss of low-valence meso-domains and the creation of high-valence meso-domains results from the loss of contacts during a deviatoric loading [SAT 78].

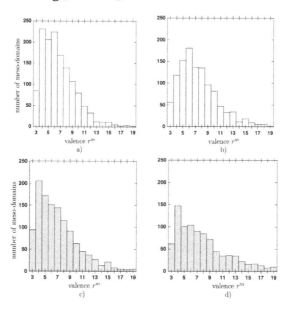

Figure 2.18. *Composition of the particle graphs constructed for samples C (plots a and b) and D (plots c and d) at the initial state (plots a and c) and at the peak state (plots b and d)*

In the current study, the *meso-scale* (or the *mesoscopic scale) for a 2D granular assembly is defined at the level of meso-domains in the corresponding particle graph*. At this scale, local texture will be described, and meso-stress and meso-strain will be defined in Chapter 3.

2.4. Some proposals for extension to 3D granular media

Some authors have considered a meso-scale in their multi-scale approach for 3D granular assemblies. In the approach of

[SAB 97], the Delaunay tessellation is used to subdivide a 3D granular medium into tetrahedra whose vertices are occupied by four particle centers. Figure 2.20 illustrates a tetrahedron obtained from this tessellation. The meso-scale defined in this approach corresponds to the level of individual tetrahedra.

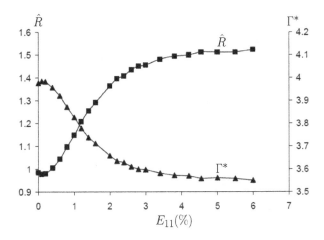

Figure 2.19. *The mean redundancy number, $\hat{R}$, and the modified coordination number, Γ^*, versus the axial strain, E_{11}, for sample A*

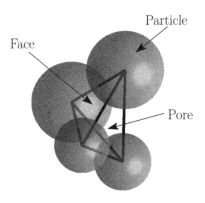

Figure 2.20. *A tetrahedron in the Delaunay tessellation*

The Delaunay tessellation has been also used by [DUR 10a, DUR 10b] to analyze the deformation characteristics of 3D granular materials from a micro-mechanical point of view. As stated by [AL 03], the Delaunay tessellation tends to artificially subdivide the void space into many void spaces, and a more realistic subdivision of the pore space can be obtained by merging certain tetrahedra using a merging criterion. We follow this technique for subdividing a 3D granular assembly into meso-domains. In the following, some merging criteria are presented.

2.4.1. *Merging criterion based on contacts between particles*

As illustrated in Figure 2.20, a tetrahedron has four triangular faces, each of which connects three particle centers. In this merging criterion, a face is said to be *closed* if it connects three particles in contact with each other, and otherwise to be *open* (Figure 2.21). Every two tetrahedra sharing an open face are merged together. Using this merging criterion, a 3D granular assembly is subdivided into a few meso-domains, some of which occupy more than thirty percent of the volume of the assembly. Therefore, this criterion is not suited for merging tetrahedra.

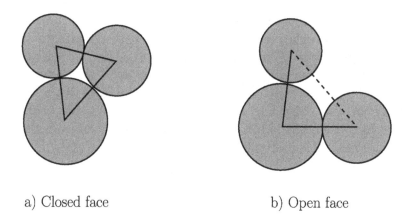

a) Closed face   b) Open face

Figure 2.21. *Illustration of a) a closed face and b) an open face*

2.4.2. *Merging criterion based on the largest void sphere*

[RAU 03] proposed a merging criterion based on the concept of the largest sphere which is entirely contained in the void space and inscribed between four particles at the vertices of each tetrahedron (the solid sphere in Figure 2.22). This largest sphere is obtained using a numerical optimization tool presented in [RAU 03].

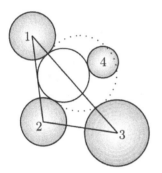

... Largest sphere which can be inscribed between the particles vertices of the Delaunay cell.

—— **The inscribed sphere** (entirely contained in the void space).

Figure 2.22. *Illustration in 2D of the largest sphere which is entirely contained in the void space and inscribed between particles at the vertices of each Delaunay cell [REB 08]*

If two largest spheres attached to the two adjacent tetrahedra overlap each other, these two tetrahedra are merged together (Figure 2.23). [REB 08] defined a merging level. For the merging level 1, only two nearest neighbors are merged together. For the merging level 2, not only the nearest neighbors but also the next nearest neighbors are merged together. [REB 08] found that this merging criterion leads to a more realistic subdivision of void space.

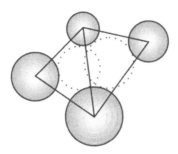

Figure 2.23. *Two tetrahedra are merged together when the two largest spheres attached to them overlap each other [REB 08]*

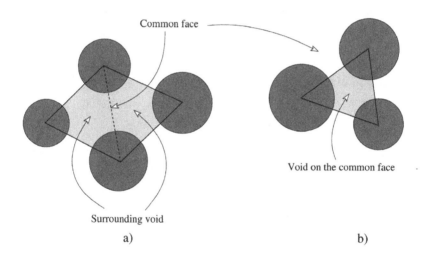

Figure 2.24. *a) Illustration in 2D of two adjacent tetrahedra sharing b) a common triangular face*

2.4.3. *Merging criterion based on the ratio of the porosity of a face to the porosity in the vicinity*

Two adjacent tetrahedra share a common triangular face (Figure 2.24). This common face has a solid fraction and a void fraction. Therefore, a porosity ϕ_f can be defined for it. The volume around the common face is defined as the union of the two tetrahedra sharing this face, and a porosity ϕ_v can be defined for this surrounding volume. According to this merging criterion, a face is considered to be *closed* if its porosity ϕ_f is small enough compared to the surrounding porosity ϕ_v, that is to say $\phi_f/\phi_v \leq \varsigma$, and otherwise is considered to be open. Every two tetrahedra sharing an open common face are merged together. Consequently, this merging criterion makes use of a threshold ς which is arbitrarily chosen. A first analysis has been carried out by [NGU 09a] to determine a reasonable value of the threshold ς. When the threshold ς decreases, the percentage $\mathfrak{p}$ of closed faces among the triangular faces also decreases. A correlation between these two parameters, ς and $\mathfrak{p}$, has been calculated. Then, by varying the value of $\mathfrak{p}$, it was found that $\mathfrak{p} = 70\%$ gives a good partition of 3D granular assemblies into meso-domains. More research on this point, however, needs to be done to obtain a more scientifically sound threshold.

3

Texture, Stress and Strain
at the Meso-scale

In this chapter, we will describe the internal state at the
meso-scale and define stress and strain tensors at this scale.
The heterogeneity of the internal state, of stress and of strain
at this scale is also presented.

3.1. Description of the internal state at the meso-scale

The internal state of granular materials can be described
by two kinds of variables: (1) scalar characteristics, which are
used for the description of compactness and (2) tensorial
variables, which are used for the description of the internal
texture, the internal anisotropy in particular. At the
considered meso-scale we consider the two kinds of variables:
(1) two scalar variables, which are the porosity and the
valence of each meso-domain and (2) a loop tensor to describe
the shape and the orientation of each meso-domain.

3.1.1. Compactness at meso-scale

In this section, we will consider two scalar variables to
describe the compactness of each meso-domain. The first one
is the porosity, denoted by ϕ^m, of each meso-domain. It is

worth remembering that all the rattlers are removed from the particle graph, hence, are not taken into account when calculating the porosity, ϕ^m. The porosity, Φ, of the sample without rattlers is the average porosity weighted by volume over all the meso-domains:

$$\Phi = \frac{1}{V} \sum_{m=1}^{N_m} \phi^m V^m, \qquad\qquad [3.1]$$

where V is the total volume of the sample and N_m is the total number of meso-domains over the sample. The cumulative distribution of the meso-domain porosity, ϕ^m, is shown in Figure 3.1. We can see that the density at the meso-scale is strongly heterogeneous and that the meso-domain porosity, ϕ^m, tends to increase during the loading due to the dilation of the sample.

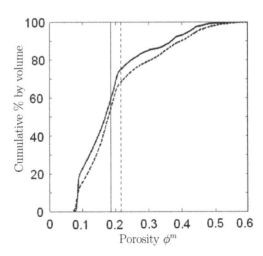

Figure 3.1. *Cumulative distribution of the meso-domain porosity, ϕ^m, at the initial state (solid line) and at $E_{11} = 5\%$ (dashed line). The two vertical lines indicate the porosity, Φ, of sample A at the two considered states*

The second scalar variable is the valence, r^m, which is the number of vertices of each meso-domain. As mentioned already in section 2.3, the meso-domain valence, r^m, is also very heterogeneous (Figure 2.17). As shown in Figure 3.2, there exists a strong correlation between the meso-domain porosity, ϕ^m and the meso-domain valence, r^m: low-valence meso-domains tend to occupy dense zones in the sample while high-valence meso-domains tend to occupy loose zones. This correlation does not hold if rattlers are taken into account when calculating the meso-domain porosity, ϕ^m, however.

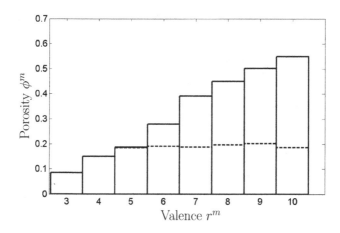

Figure 3.2. *Average value of the porosity ϕ^m for each set of meso-domains having the same valence r^m. The meso-domain porosity, ϕ^m, is calculated without (solid line) and with (dashed line) rattlers*

3.1.2. *Texture at meso-scale*

[TSU 98] have proposed a loop tensor, denoted by $\boldsymbol{F}^m$, to describe the texture of each meso-domain:

$$\boldsymbol{F}^m = \frac{1}{2}\sum_{k=1}^{r^m} \boldsymbol{l}^k \otimes \boldsymbol{l}^k, \qquad [3.2]$$

where l^k is the branch vector at contact k: $l^k = l^k n_b^k$ with length l^k and unit vector n_b^k directed along branch k (Figure 3.3) and r^m is the number of sides of meso-domain m. We have:

$$\text{Trace}(\boldsymbol{F}^m) = \frac{1}{2}\sum_{k=1}^{r^m} l^k l^k, \qquad [3.3]$$

which is a measure of the meso-domain size.

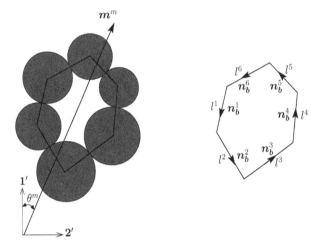

Figure 3.3. *The length l^k of each branch k and the unit vector n_b^k directed along each branch k. Each meso-domain has an elongation direction m^m and an orientation angle θ^m with respect to the compression direction, $1'$*

We have modified this loop tensor such that its trace measures the meso-domain perimeter [NGU 09b]:

$$\boldsymbol{L}^m = \sum_{k=1}^{r^m} l^k n_b^k \otimes n_b^k. \qquad [3.4]$$

It is easy to find that $\text{Trace}(\boldsymbol{L}^m) = \sum_{k=1}^{r^m} l^k$ and the loop tensor, $\boldsymbol{L}^m$, is symmetric. It should be noted that the unit vector n_b^k directed along each branch k is different from the unit normal vector n^k at each contact k in general case. Nevertheless, these two unit vectors coincide for an assembly of circular particles.

The loop tensor, $\boldsymbol{L}^m$, describes the texture of each meso-domain in terms of the orientation and of the length of branches joining two particles in contact. Note that the tensor $\boldsymbol{L}^m$ for a regular polygon is isotropic. Summing tensors $\boldsymbol{L}^m$ over all meso-domains gives a modified fabric tensor, $\boldsymbol{H}^*$, which takes into account the size of particles in contact:

$$\boldsymbol{H}^* = \frac{1}{2} \sum_{m=1}^{N_m} \boldsymbol{L}^m = \sum_{k=1}^{N_c} l^k n_b^k \otimes n_b^k, \qquad [3.5]$$

where fraction $1/2$ indicates that two meso-domains share a branch. For an assembly of circular particles, the tensor $\boldsymbol{H}^*$ coincides with the modified fabric tensor used by [NEM 83] to describe the texture of granular assemblies.

Loop tensor, $\boldsymbol{L}^m$, for each meso-domain is decomposed into an isotropic part and a deviatoric part as follows:

$$\boldsymbol{L}^m = \frac{1}{2}\text{Trace}(\boldsymbol{L}^m)\boldsymbol{I} + \boldsymbol{D}^m. \qquad [3.6]$$

The deviatoric part $\boldsymbol{D}^m$ characterizes the shape and the orientation of the meso-domain under consideration. It gives two important pieces of information about the geometry of the meso-domain:

1) The *elongation degree*, β^m, of the meso-domain m is defined as:

$$\beta^m = \frac{\|\boldsymbol{D}^m\|}{\text{Trace}(\boldsymbol{L}^m)} = \sqrt{\frac{1}{2}} \frac{L_1^m - L_2^m}{L_1^m + L_2^m}, \qquad [3.7]$$

where L_1^m and L_2^m are the major and minor eigenvalues of L^m, respectively.

2) The *elongation direction*, m^m, of the meso-domain m corresponds to the major principal direction of D^m. The orientation angle, θ^m, of the meso-domain m is defined as the angle between the elongation direction, m^m and the compression direction, $1'$ (Figure 3.3).

The elongation degree, β^m, is zero for a meso-domain of regular shape; and the higher the elongation degree, β^m, is, the more the meso-domain is elongated. Figure 3.4 presents six meso-domains with different elongation degrees, β^m, from 0 to 0.5 and with different orientations, θ^m.

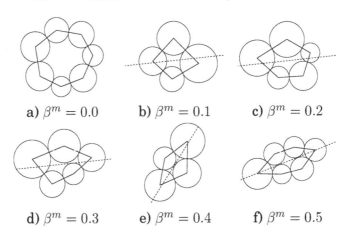

a) $\beta^m = 0.0$ b) $\beta^m = 0.1$ c) $\beta^m = 0.2$

d) $\beta^m = 0.3$ e) $\beta^m = 0.4$ f) $\beta^m = 0.5$

Figure 3.4. *Meso-domains with different elongation degrees, β^m, and different orientations, θ^m, given by the dashed lines*

The range of variation of the elongation degree, β^m, for each set of meso-domains having the same valence, r^m, is presented in Figure 3.5. We can see that triangular meso-domains have low elongation degrees. On the other hand, the elongation degree for other meso-domains varies widely from 0 up to 0.5. The cumulative distribution of the meso-domain elongation degree is shown in Figure 3.6. The meso-domain elongation degree is very heterogeneous and it changes slightly during the loading.

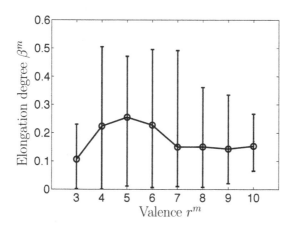

Figure 3.5. *Bars showing the variation of the elongation degree, β^m, for each set of meso-domains having the same valence, r^m. The marker (○) indicates the mean value of β^m for each set*

It can, thus, be identified that loop tensor, L^m, gives complete information on a given meso-domain, the size (perimeter given by its trace), the orientation, θ^m, and the shape (elongation degree, β^m), in particular.

3.1.3. *Classification of meso-domains*

Meso-domains are classified into two elongation classes and then into three orientation sub-classes depending on their elongation degree, β^m and their orientation, θ^m. This classification is needed to facilitate the analysis of the effect of the texture of meso-domains on their behavior. For this purpose, all the meso-domains are split into two elongation classes initially, W and S, containing *weakly* (class W) and *strongly* (class S) elongated meso-domains, respectively. A meso-domain is considered weakly elongated if its elongation degree, β^m, is smaller than the threshold value, and strongly elongated otherwise. In our study, the threshold value was chosen to be equal to the volumetric average elongation degree, $\bar{\beta}$, over all meso-domains at the initial state. For example, the average elongation degree, $\bar{\beta}$, at the initial state is 0.2 for sample A, which is chosen as the threshold value to classify meso-domains for this sample. It can be noted that the average elongation degree, $\bar{\beta}$, for samples A and B changes slightly during loading.

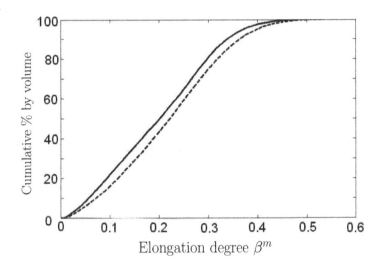

Figure 3.6. *Cumulative distribution of the meso-domain elongation degree, β^m, at the initial state (solid line) and at $E_{11} = 5\%$ (dashed line)*

The meso-domains in each elongation class are next split into three sub-classes $1'$, $0'$ and $2'$ containing meso-domains oriented in the compression direction, $1'$, obliquely and in the extension direction, $2'$, respectively (Figure 3.7).

In doing so, all the meso-domains are split into six sub-classes: $W1'$, $W0'$ and $W2'$ for class W and $S1'$, $S0'$ and $S2'$ for class S. In each sub-class, the first character (W or S)

stands for the meso-domain elongation degree and the second one ($1'$, $0'$ or $2'$) stands for the meso-domain orientation. These six sub-classes of meso-domains can be thought of as being *six local phases* of the sample, and their texture represents the local internal state of the sample. The number of phases (six for this study) is arbitrary. In our opinion, it is the smallest number that can be considered, because it is necessary to take into account the meso-domain elongation degree (at least two elongation classes: W and S) and the meso-domain orientation with respect to the loading direction (at least three orientation sub-classes have to be considered to clearly identify the role of the anisotropy in the behavior of the material). It is possible to consider a higher number of phases to get more accurate descriptions, but the analysis would be more complex.

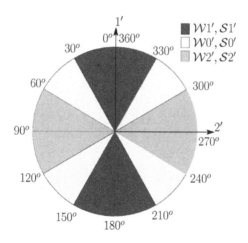

Figure 3.7. *Classification of the meso-domains into six sub-classes $W1'$, $W0'$ and $W2'$ for class W and $S1'$, $S0'$ and $S1'$ for class S depending on the orientation angle, θ^m, of each meso-domain. Each sub-classe is afterward called a phase*

Figure 3.8 shows the spatial distribution of the six phases throughout the loading path performed on sample B. It is shown that there is no spatial localization of any phases during the loading. This result justifies the use of the six defined phases as basic units to characterize the meso-scale and as a framework for a classic change of scale using the volume average for the entire sample.

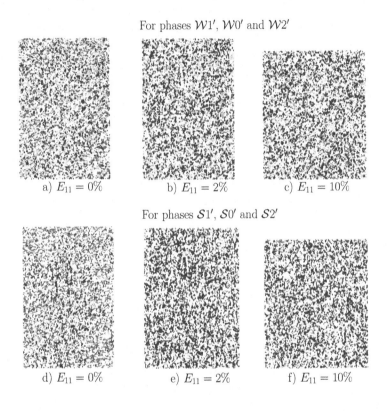

For phases $\mathcal{W}1'$, $\mathcal{W}0'$ and $\mathcal{W}2'$

a) $E_{11} = 0\%$ b) $E_{11} = 2\%$ c) $E_{11} = 10\%$

For phases $\mathcal{S}1'$, $\mathcal{S}0'$ and $\mathcal{S}2'$

d) $E_{11} = 0\%$ e) $E_{11} = 2\%$ f) $E_{11} = 10\%$

Figure 3.8. *Spatial distribution of the six phases $\mathcal{W}1'$, $\mathcal{S}1'$ (blue), $\mathcal{W}0'$, $\mathcal{S}0'$ (green), and $\mathcal{W}2'$ and $\mathcal{S}2'$ (red) defined for sample B at $E_{11} = 0\%$, 2%, and 10% during the performed loading test. For a color version of this figure, see www.iste.co.uk/cambou/mesoscale.zip*

3.2. Definition of strain at the meso-scale

As presented previously, a granular assembly can be sub-divided into polygonal meso-domains, each of which can be, in turn, sub-divided into triangles. A natural way to define a strain tensor for each meso-domain is to consider the volume-weighted average of strain tensors defined for each triangle belonging to the meso-domain in question. Another method, which is more sophisticated, is to assign a continuous displacement field to each meso-domain, and to then calculate the volume-weighted average of the assigned displacement gradient. This section deals with the presentation of the first approach. The reader can refer to the paper of [KRU 96] for the second approach. In the following section a strain tensor is first defined for each triangle and an averaged strain tensor is then derived for each meso-domain.

3.2.1. Strain tensor for each triangle of the Delaunay tesselation

For this definition, an incremental strain tensor, $\delta\varepsilon^e$, is first defined for each triangle. [CAM 00] assumed that the strain field is constant in each triangular element, e. The symbol δ indicates that the strain tensor, $\delta\varepsilon^e$, is defined for a small loading increment. This approach has been presented in section 1.3.1. According to this approach, the displacement gradient tensor, δa^e, for each triangular element is given by equation [1.30] which can be rewritten as:

$$\delta a^e = \delta u^{10} \otimes y^{10} + \delta u^{20} \otimes \Delta y^{20}. \qquad [3.8]$$

Three vertices 0, 1 and 2 of each triangular element are defined in counterclockwise order as illustrated in Figure 3.9. In equation [3.8], vertex 0 is chosen as the reference vertex. The two vectors δu^{10} and δu^{20} are relative displacements of the two vertices 1 and 2 with respect to the reference vertex, 0 ($\delta u^{v0} = \delta u^v - \delta u^0$, $v = 1, 2$). The two vectors y^{10} and y^{20} are defined such that:

$$x^{10} \otimes y^{10} + x^{20} \otimes y^{20} = I, \qquad [3.9]$$

where $x^{v0} = x^v - x^0$, $v = 1,\ 2$ is the branch vector connecting vertex v to the reference vertex 0.

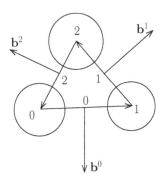

Figure 3.9. *A triangle whose vertices are numbered 0, 1 and 2 and whose sides are numbered 0, 1 and 2. Three vectors b^0, b^1 and b^2 are associated with the three sides 0, 1 and 2, respectively and are used in equation* [3.16]

Solving the system of four linear independent equations in [3.9], we obtain:

$$y_1^{10} = \frac{1}{2V^e}x_2^{20}, \quad y_2^{10} = -\frac{1}{2V^e}x_1^{20}$$

$$y_1^{20} = -\frac{1}{2V^e}x_2^{10}, \quad y_2^{20} = \frac{1}{2V^e}x_1^{10},$$

[3.10]

where V^e is the area of the considered triangle e. Let b^{wv} (w and v are two arbitrary vertices) be the vector obtained by rotating the vector x^{wv} by 90° in such a way that it points outward. We have $b^{10} = [x_2^{10}, \; -x_1^{10}]$ and $b^{20} = [-x_2^{20}, \; x_1^{20}]$. Equation [3.11] can be then rewritten as:

$$y^{10} = -\frac{b^{20}}{2V^e} \quad \text{and} \quad y^{20} = -\frac{b^{10}}{2V^e}.$$

[3.11]

Combining equations [3.8] and [3.11], the displacement gradient tensor, δa^e, is expressed as:

$$\delta a^e = -\frac{1}{2V^e}\Big(\delta u^{10} \otimes b^{20} + \delta u^{20} \otimes b^{10}\Big).$$

[3.12]

It should be noted that the displacement gradient tensor, δa^e, does not depend on the choice of the reference vertex. Choosing the vertices 1 and 2, in turn, as the reference vertex gives:

$$\delta a^e = -\frac{1}{2V^e}\left(\delta u^{01} \otimes b^{21} + \delta u^{21} \otimes b^{01}\right), \qquad [3.13]$$

$$\delta a^e = -\frac{1}{2V^e}\left(\delta u^{02} \otimes b^{12} + \delta u^{12} \otimes b^{02}\right). \qquad [3.14]$$

Note that $\delta u^{wv} = -\delta u^{vw}$ and $b^{wv} = b^{vw}$. By summing equations [3.12]–[3.14], we obtain:

$$\delta a^e = \frac{1}{6V^e}\left[\delta u^{01} \otimes (b^{12} - b^{20}) + \delta u^{12} \otimes (b^{20} - b^{01}) \quad [3.15]\right.$$
$$\left. + \delta u^{20}(b^{01} - b^{12})\right].$$

Three sides 0-1, 1-2 and 2-0 of each triangular element are also numbered 0, 1 and 2 in counterclockwise order as illustrated in Figure 3.9. For a side s going from vertex w to vertex v, we introduce δu^s and b^s such that $\delta u^s = \delta u^{wv}$ and $b^s = b^{wv}$, with $s \in 0, 1, 2$. Equation [3.15] can be rewritten as:

$$\delta a^e = \frac{1}{6V^e} \sum_{s_1,s_2 \in \{0,1,2\}} Q^3_{s_1,s_2}\delta u^{s_1} \otimes b^{s_2}. \qquad [3.16]$$

Matrix Q^3 has been introduced by [KUH 97]:

$$Q^3 = \begin{bmatrix} 0 & 1 & -1 \\ -1 & 0 & 1 \\ 1 & -1 & 0 \end{bmatrix}. \qquad [3.17]$$

The two superscripts s_1 and s_2 denote two sides of triangle e with values (0, 1, 2).

The strain tensor, $\delta\varepsilon^e$, is the symmetric part of the displacement gradient tensor, δa^e:

$$\delta\varepsilon^e = \frac{1}{2}\left(\delta a^e + \delta a^e\right). \qquad [3.18]$$

[BAG 96] proposed another method for defining an average strain tensor for each triangle by assigning a displacement field $\delta u(x)$ to it. The displacement at each vertex is equal to that of the particle center corresponding to the vertex; and the displacement along each side is linearly interpolated from that of its vertices. It should also be noted that the displacement of points inside the triangle does not affect the obtained strain tensor since the volume integral of the displacement gradient is transformed to the surface integral using Gauss-Ostrogradski Theorem. The strain tensor given by this method is identical to the one obtained by the method presented above. This is not surprising since when the constant strain is assumed, the displacement field is linear along each side, which satisfies Bagi's displacement field [BAG 96].

Figure 3.10 shows the spatial distribution of the strain component $\delta\varepsilon_{11}^e$ defined for each triangular element at different loading levels during the biaxial test performed on sample D composed of octagonal particles. This figure highlights a spatial heterogeneity of the strain field, $\delta\varepsilon^e$, defined for each triangular element. At the beginning of loading (E_{11} = 0.3%), the strain is quite homogeneous. However, the strain becomes heterogeneous as the axial strain E_{11} increases. At E_{11} = 2.5% (the peak state), several shear bands become apparent and they persist until the end of the test (E_{11} = 8.0%).

3.2.2. Strain tensor for each meso-domain

Once the displacement gradient tensor, δa^e, is defined for each triangle, the displacement gradient tensor, δa^m, for a given meso-domain m can be defined as the average volume over all the triangles included in this meso-domain (Figure 3.11):

$$\delta \boldsymbol{a}^m = \frac{1}{V^m} \sum_{e=1}^{N_e^m} \delta \boldsymbol{a}^e V^e,$$ [3.19]

where N_e^m is the number of triangles included in meso-domain m.

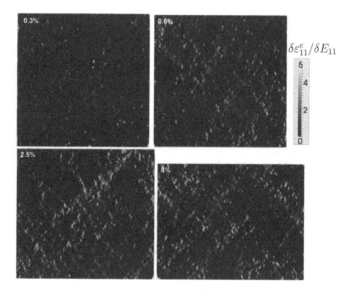

Figure 3.10. *Spatial distribution of the strain component $\delta \varepsilon_{11}^e$ defined for each triangular element at different values of the macroscopic axial strain, E_{11}, for sample D. The color scale corresponds to the value of $\delta \varepsilon_{11}^e$ normalized by the macroscopic axial strain increment, δE_{11}. For a color version of this figure, see www.iste.co.uk/cambou/mesoscale.zip*

Based on the same approach, [KUH 97] derived an expression of the average strain tensor for each polygonal meso-domain as follows:

$$\delta \boldsymbol{a}^m = \frac{1}{6V^m} \sum_{s_1,s_2 \in \{0,1,\dots,r^m-1\}} Q_{s_1,s_2}^{r^m} \delta \boldsymbol{u}^{s_1} \otimes \boldsymbol{b}^{s_2}.$$ [3.20]

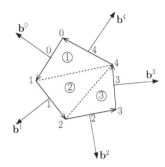

Figure 3.11. *Vertices, sides and triangles of a meso-domain*

The index s and the vector b^s for each side are illustrated in Figure 3.11. The matrix Q^{r^m} of dimension $r^m \times r^m$ is skew-symmetric and circulant ($Q^{r^m}_{s_1,s_2} = Q^{r^m}_{s_1-1,s_2-1}$). For a triangle ($r^m = 3$), the matrix Q^3 is given by equation [3.17]. First row of the matrix Q^{r^m} for r^m = 4, 5 and 6 are given by Table 3.1. The remaining rows can be obtained by using the circular form of Q^{r^m}.

Valence r^m	s_2					
	0	1	2	3	4	5
4	0	3/2	0	3/2		
5	0	9/5	3/5	-3/5	-3/5	
6	0	2	1	0	-1	-2

Table 3.1. *First row of matrix Q^{r^m} for different values of the valence, r^m*

Equation [3.20] shows that the displacement gradient tensor, δa^m, defined in equation [3.19] does not depend on the way the considered meso-domain is triangulated. The strain tensor, $\delta \varepsilon^m$, for each meso-domain is the symmetric part of the displacement gradient tensor, δa^m. [KRU 96] have derived another expression of the strain tensor, $\delta \varepsilon^m$, for each meso-domain by using the same assumption as in [BAG 96] presented in section 3.2.1. The obtained expression for $\delta \varepsilon^m$ is different from equation [3.20]; however the value of $\delta \varepsilon^m$ is exactly the same.

The increment of the global strain tensor of the sample can be calculated by the volume average of the local one, $\delta\varepsilon^m$, over all the meso-domains:

$$\delta\bar{\varepsilon} = \frac{1}{V}\sum_{m=1}^{N_m}\delta\varepsilon^m V^m. \qquad [3.21]$$

The increment of the strain tensor, $\delta\varepsilon^m$, for each meso-domain as well as the average of the increment of the strain tensor, $\delta\bar{\varepsilon}$, for samples A and B are calculated for each axial strain increment $\delta E_{11} = 0.02\%$ prescribed at different values of the macroscopic axial strain, E_{11}. It is worth mentioning that the increment of the strain tensors $\delta\varepsilon^m$ and $\delta\bar{\varepsilon}$ depend on whether the initial configuration or the final configuration is considered for the construction of the particle graph. However, this dependence is negligible if the prescribed strain increment is sufficiently small. It was found that the axial strain increment $\delta E_{11} = 0.02\%$ chosen in our calculation leads to a very small difference between the average of the increment of the strain tensor, $\delta\bar{\varepsilon}$, calculated with the initial configuration and the one with the final configuration. Therefore, the initial configuration of each increment is considered.

Figure 3.12 shows a comparison between the increment of the average strain tensor, $\delta\bar{\varepsilon}$, estimated by equation [3.21] and the increment of the strain tensor, δE, defined from the boundary conditions for different strain increments prescribed to sample A at different values of the axial strain, E_{11}. It can be seen that the estimated increment of the strain tensor, $\delta\bar{\varepsilon}$, for the sample from its homologue defined for each meso-domain, is very close to the increment of the strain tensor, δE, defined from the boundary conditions. This mean that the meso-scale is relevant for a change of scale in terms of kinematic variables.

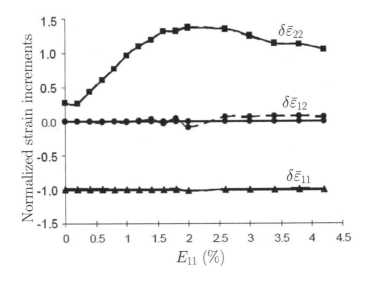

Figure 3.12. *Increment of the average strain tensor, $\delta\bar{\varepsilon}$, (dotted line) estimated by equation [3.21] compared to the increment of the macroscopic strain tensor, δE, defined for sample A (solid line) at different loading levels. Both tensors $\delta\bar{\varepsilon}$ and δE are normalized by the prescribed increment of axial strain $\delta E_{11} = 0.02\%$*

Figure 3.13 presents the spatial distribution of the increment of the volumetric strain, $\delta\varepsilon_v^m = \delta\varepsilon_{11}^m + \delta\varepsilon_{22}^m$, defined for each meso-domain at different loading levels performed on sample A. Each meso-domain is colored according to its increment of volumetric strain, $\delta\varepsilon_v^m$, normalized by the axial strain increment, δE_{11}, as prescribed to the sample. It can be seen that during the beginning of the loading, the increment of strain is more or less homogeneous at the meso-scale and almost all meso-domains contract. With the increasing deviatoric loading, strain becomes more and more heterogeneous. In the same sample, some meso-domains contract, while the others dilate and the number of dilatant meso-domains increases with the loading. At the axial strain $E_{11} = 2.0\%$, which corresponds to the peak of the stress-strain curve shown in Figure 2.4, strain increments are localized strongly within one shear band within which almost all meso-domains dilate strongly. This shear band is replaced afterward by two new ones as observed at the axial strain $E_{11} = 3.0\%$.

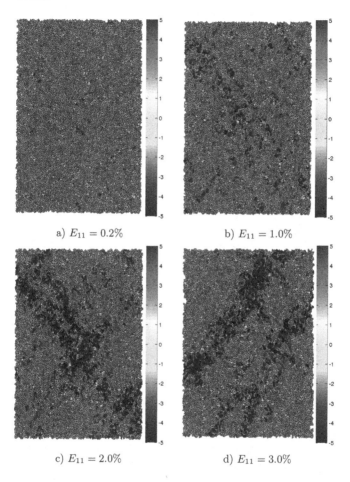

a) $E_{11} = 0.2\%$ b) $E_{11} = 1.0\%$

c) $E_{11} = 2.0\%$ d) $E_{11} = 3.0\%$

Figure 3.13. *Spatial distribution of the volumetric strain, $\delta\varepsilon_v^m$, defined for each domain at different values of the macroscopic axial strain, E_{11}. The color scale corresponds to the value of $\delta\varepsilon_v^m$ normalized by the prescribed axial strain increment $\delta E_{11} = 0.02\%$. For a color version of this figure, see www.iste.co.uk/cambou/mesoscale.zip*

3.3. Definition of stress at the meso-scale

The definition of stress at the meso-scale is still an open issue. The main difficulty in defining this meso-stress field is that the volume, V^m, of a meso-domain in the particle graph does not match the volume, $\hat{V}^m$, to which the external forces are applied (Figure 3.14). The volume, $\hat{V}^m$, is defined by addition of all the Dirichlet polygonal cells associated with all the particles included in the considered meso-domain m. Forces in contact between the particles of a given meso-domain and the neighboring particles in contact are acting on the boundary of $\hat{V}^m$, so $\hat{V}^m$ appears to be a correct reference volume to compute meso-stress. Nevertheless, this volume $\hat{V}^m$ has the drawback of being non-additive, i.e. volumes $\hat{V}^m$ intersect each other.

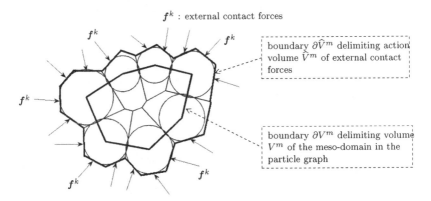

f^k : external contact forces

boundary $\partial \hat{V}^m$ delimiting action volume $\hat{V}^m$ of external contact forces

boundary ∂V^m delimiting volume V^m of the meso-domain in the particle graph

Figure 3.14. *Illustration of a meso-domain, whose volume in the particle graph is defined by V^m, loaded by the external contact forces applied on an external boundary, $\partial \hat{V}^m$, delimiting a greater volume, $\hat{V}^m$. The volume $\hat{V}^m$ is composed of the Dirichlet hexagonal cells associated with the particles belonging to meso-domain m*

One can find in the literature some definitions of a stress field in an assembly of particles, which can be extended easily to the case of meso-domains. For example, [SAB 97] has proposed a method to define a stress tensor for each tetrahedron joining four particle centers in a 3D granular material (particles not necessarily in contact), as follows:

$$\sigma^m = \sum_{k=1}^{6} \frac{1}{\tilde{V}^k} f^k \otimes^s l^k, \qquad [3.22]$$

where f^k and l^k are the contact force and the branch vector of edge k, $\tilde{V}^k$ is the total volume of all the tetrahedra sharing edge k and symbol $\otimes^s$ denotes the symmetric tensor product: $u \otimes^s v = (u_i v_j + u_j v_i)/2$. The extension of this definition to the case of 2D meso-domain is straightforward:

$$\sigma^m = \sum_{k=1}^{r^m} \frac{1}{\tilde{V}^k} f^k \otimes^s l^k, \qquad [3.23]$$

where r^m is the valence of the considered meso-domain m and $\tilde{V}^k$ is the total surface of the two meso-domains sharing edge k. This definition thus consists of distributing the force at a given contact to all the meso-domains sharing this contact, by weighting the volume of these meso-domains. It should be noted that relations [3.22] and [3.23] are derived directly from equation [1.14] by associating a local volume to each contact, leading to two main drawbacks of this approach. Firstly, the stress tensor given by relation [1.14] is not symmetric for a collection of a number of particles. This is the case of meso-domains, so the symmetrization operation ($\otimes^s$) must be used in equations [3.22] and [3.23] to obtain symmetric stress tensors at the meso-scale. Secondly, it is difficult to propose a clear definition of the local volume, $\tilde{V}^k$, associated to each contact.

[EHL 03] have proposed a method for defining an average stress tensor for a local volume that consists of an ensemble of particles, whose boundary passes through the mass centers of the outermost particles. This local volume is therefore similar to meso-domains. Extending this method to each meso-domain gives:

$$\sigma^m = \frac{1}{V^m} \sum_{g=1}^{r^m} f^g \otimes x^g, \qquad\qquad [3.24]$$

where g runs over all the particles of the meso-domain m, f^g is the resultant external force acting on particle g and x^g is the position of the mass center of particle g. The main advantage of this definition is that the stress tensor, σ^m, for a given meso-domain is directly defined on its volume, V^m, from its external contact forces. Nevertheless, the major drawback is that the stress tensor obtained from this approach is not symmetric.

To overcome these difficulties, we have proposed three methods in [NGU 09b, NGU 12] to define the meso-stress tensor, two of which are presented in this book as they possess a clear physical basis.

3.3.1. Method I: meso-stress defined by assigning a local stress field to each material point

3.3.1.1. Description of method I

This method assumes that a granular medium can be replaced by an equivalent continuum in which a local stress field is assigned to each material point. This local stress field is denoted by $\sigma(x)$, where x is the position vector of a point in the continuum. In this way, each meso-domain occupies a finite volume, V^m, in the equivalent continuum and an average stress tensor, σ^m, can be consistently defined (in the sense of continuum mechanics) over the volume V^m by:

$$\sigma^m = \frac{1}{V^m} \int_{V^m} \sigma(x) dV. \qquad\qquad [3.25]$$

The key point of this first method is how to assign a local stress field, $\sigma(x)$, to the equivalent continuum. In [NGU 12], we adopted the method proposed by [SAT 92, SAT 04] and [KUH 03], which consists of filling a granular medium with Dirichlet cells, each of which is associated to a particular particle (see section 2.3 for definition of a Dirichlet cell and Figure 2.12 for the illustration of a Dirichlet system). One Dirichlet cell thus contains a given particle and a portion of the void around this particle. Then, given a particle g and its Dirichlet cell, D^g, an average stress tensor in this cell, σ^g, is usually defined from the external contact forces exerted on it [AUV 92, BAC 92, BAG 96, KUH 03]:

$$\sigma^g = \frac{1}{V_D^g} \sum_{k=1}^{N_c^g} f^k \otimes r^k, \qquad [3.26]$$

where V_D^g is the volume of the associated Dirichlet cell, D^g, N_c^g is the number of particle contacts g, f^k is the external force exerted on particle g at the contact point k and r^k is the contact vector joining the particle center to the contact point. As shown in Figure 3.15, this field is by definition constant on each Dirichlet cell. Moreover, this field is heterogeneous within each meso-domain (black lines in Figure 3.15) and discontinuous through the boundaries between Dirichlet cells, which also implies a discontinuous stress vector on these boundaries. As a consequence, the static balance is not fulfilled at the boundaries between Dirichlet cells.

Once the local stress field, $\sigma(x)$, is defined at each material point in the equivalent continuum, stress tensor, σ^m can now be defined for each meso-domain using equation [3.25]. It follows that:

$$\sigma^m = \frac{1}{V^m} \sum_{g=1}^{r^m} \sigma^g \Delta V_D^g, \qquad [3.27]$$

where g runs over all the particles of meso-domain m, σ^g is the average stress tensor defined by equation [3.26] on the

Dirichlet cell associated to particle g and ΔV_D^g is the volume fraction of the Dirichlet cell inside meso-domain m. Equation [3.27] shows that the stress tensor, σ^m, defined for each meso-domain, is the volume-average stress over all the portions of the Dirichlet cells associated to the considered meso-domain. This definition is also in agreement with that proposed by [LÄT 00], which computes the average stress tensor for any volume in a granular material from the local stress field classically defined for each particle.

Figure 3.15. *Spatial distribution of the mean stress, p, in the equivalent continuum at the initial state. The assigned stress tensor, σ, is constant over each Dirichlet cell and is given by equation [3.26]. Gray levels represent the value of the ratio p/P, where P is the macroscopic mean stress. Solid lines correspond to closed branches delimiting meso-domains in the particle graph*

3.3.1.2. *Advantage and drawbacks of method I*

The major advantage of this first method is that the average stress tensor, σ^m, for a meso-domain is defined on its volume, V^m, in the particle graph. Nevertheless, this method also presents two important drawbacks:

– the local stress field, $\sigma(x)$, defined at each point in the continuum medium is assumed to be constant in each Dirichlet cell, which obviously is not the case;

– the local stress field, $\sigma(x)$, does not ensure the static equilibrium of meso-domains. Let us denote by $f^m(x)$ the stress vector associated to this local stress field, $\sigma(x)$, acting on the boundaries of a meso-domain m:

$$f^m(x) = \sigma(x).n^m(x),$$ [3.28]

where $n^m(x)$ is the unit outward normal vector to the considered meso-domain at a point x on the boundaries. It is then easy to verify that the equilibrium condition is not ensured:

$$\int_{\partial V^m} f^m(x)dS = \int_{\partial V^m} \sigma(x).n^m(x)dS \neq 0.$$ [3.29]

3.3.2. Method II: meso-stress defined from external contact forces applied to the particles of each meso-domain

3.3.2.1. Description of method II

As presented in section 3.2, the strain tensor, $\delta\varepsilon^m$, of a meso-domain m is consistently defined from displacements of the centers of its particles. These displacements are themselves induced by the contact forces applied to the boundary, $\partial\hat{V}^m$, of the considered meso-domain (Figure 3.14). Hence, an appropriate dual variable of the meso-strain would be a local stress tensor defined at the same level from external contact forces applied to each meso-domain. An average stress tensor can be consistently defined on the volume $\hat{V}^m$ using the method proposed in [CHR 81]:

$$\hat{\sigma}^m = \frac{1}{\hat{V}^m} \int_{\hat{V}^m} \sigma(x)dV.$$ [3.30]

Using Gauss's theorem and the local static balance condition $\operatorname{div}\sigma(x) = 0$, equation [3.30] can be rewritten in a discrete form as follows [MAG 08]:

$$\hat{\sigma}^m = \frac{1}{\hat{V}^m} \sum_{k \in K_{ext}^m} f^k \otimes x^k,$$ [3.31]

where K_{ext}^m is the set of external contacts of meso-domain m as shown in Figure 3.14, x^k is the position vector of contact k and f^k is the contact force at contact k. If static equilibrium is satisfied for all the particles of the considered meso-domain, then the local stress tensor defined by equation [3.31] is symmetric. This stress tensor can also be defined by a volume-average of stress tensors, σ^g, defined for each Dirichlet cell associated with the considered meso-domain:

$$\hat{\sigma}^m = \frac{1}{\hat{V}^m} \sum_{g=1}^{r^m} \sigma^g V_D^g, \qquad [3.32]$$

where σ^g is the constant stress tensor on each Dirichlet cell defined by equation [3.26].

Whatever equation ([3.31] or [3.32]) is used, it is worth mentioning that we do not use any assumption to define the average stress tensor, $\hat{\sigma}^m$, over the volume $\hat{V}^m$. However, as mentioned above, the main drawback of the set of volumes $\hat{V}^m$ is its non-additivity because of the inherent intersection of volumes $\hat{V}^m$. This leads to non-unicity of the local stress tensor at the intersection between two adjacent meso-domains. To overcome this difficulty, we assume that *the average stress tensor defined over the volume V^m is equal to the average stress tensor defined over the volume $\hat{V}^m$, that is $\sigma^m = \hat{\sigma}^m$.*

We thus have:

$$\sigma^m = \frac{1}{\hat{V}^m} \sum_{k \in K_{ext}^m} f^k \otimes x^k = \frac{\eta^m}{V^m} \sum_{k \in K_{ext}^m} f^k \otimes x^k, \qquad [3.33]$$

where $\eta^m = V^m/\hat{V}^m$ is the ratio of the volume V^m to the volume $\hat{V}^m$, which is, by definition, lower than 1.

3.3.2.2. *Advantage and drawbacks of method II*

A major advantage of this second method is that it allows a relevant local stress tensor to be defined for the volume associated with each meso-domain (the one to which external forces on each meso-domain are applied). Moreover, this local stress tensor has correct duality with the local strain tensor defined on the same volume for each meso-domain. Nevertheless, this method presents a significant drawback when assuming that the local stress tensor in the volume $\hat{V}^m$ can be applied to the volume V^m: as later explained, this leads to an overestimation of the macroscopic stress. Nevertheless, the usual macroscopic stress ratio, Q/P, is well estimated, as explained in the following section.

3.3.3. Comparisons between the two methods

We will proceed to examine some comparisons between the two methods I and II at the meso-scale and at the macro-scale, to determine which one will be used to estimate all the stress-related quantities of our analyses. For these comparisons, sample A is considered.

3.3.3.1. Comparisons at the meso-scale

First, we analyze the principal directions and eigenvalues of the meso-stress tensor in each meso-domain defined by the two methods I and II, denoted by $[\sigma^m]^I$ and $[\sigma^m]^{II}$, respectively. Figure 3.16 shows that the two methods lead to quite similar local stress tensors. Moreover, Figure 3.16 shows that the principal directions of the local stress fields tend to coincide with those of the macroscopic stress (here the vertical and horizontal directions are the major and minor principal directions of the macroscopic stress, respectively).

Then, we proceed to quantitative comparisons by analyzing the stress components of the two meso-stress tensors $[\sigma^m]^I$ and $[\sigma^m]^{II}$. To this end, we compute ratios of stress components defined by method II to respective stress components defined by method I, for each meso-domain. Table 3.2 shows distribution characteristics of these ratios for the stress components, σ_{11}^m, σ_{22}^m and σ_{12}^m and for the stress ratio, q^m/p^m, for all the meso-domains at the peak state.

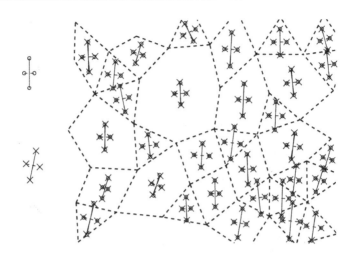

Figure 3.16. *Principal directions and eigenvalues of the local stress tensors defined in each meso-domain by methods I and II for a part of the particle graph*

Ratios	Range of variation	Mean value	Standard deviation
$[\sigma_{11}^m]^{II}/[\sigma_{11}^m]^{I}$	[0.8, 1.4]	1.04	0.06
$[\sigma_{22}^m]^{II}/[\sigma_{22}^m]^{I}$	[0.8, 1.4]	1.05	0.07
$[\sigma_{12}^m]^{II}/[\sigma_{12}^m]^{I}$	[-0.5, 2.5]	0.98	0.24
$[q^m/p^m]^{II}/[q^m/p^m]^{I}$	[0.6, 1.4]	1.0	0.08

Table 3.2. *Characteristics of the distributions of the ratios between the stress components, σ_{11}^m, σ_{22}^m, σ_{12}^m and the stress ratio, q^m/p^m, defined by methods I and II for all the meso-domains at the peak state*

It is shown that the ratios for σ_{11}^m, σ_{22}^m and q^m/p^m are gathered closely around 1.0 with small standard deviations, while the ratio for σ_{12}^m varies more widely with a standard deviation four times larger. It is worth noting that the order of magnitude of σ_{12}^m is much smaller than those of the components σ_{11}^m and σ_{22}^m during the biaxial compression test.

The above comparisons show that the two methods proposed for defining the local stress field at the meso-scale give quite similar results in terms of analyses at the meso-scale.

3.3.3.2. Comparisons at the macro-scale

We now investigate the ability of the two methods to estimate the mean stress tensor at the macro-scale, which will be an important step in the change of scale process presented in Chapter 5. Averaging local stress fields, $[\sigma^m]^I$ and $[\sigma^m]^{II}$, over all the meso-domains gives the respective approximations, $[\Sigma]^I$ and $[\Sigma]^{II}$, of the macroscopic stress tensor Σ.

Method I: $$[\Sigma]^I = \frac{1}{V} \sum_{m=1}^{N_m} [\sigma^m]^I V^m, \qquad\qquad [3.34]$$

Method II: $$[\Sigma]^{II} = \frac{1}{V} \sum_{m=1}^{N_m} [\sigma^m]^{II} V^m, \qquad\qquad [3.35]$$

where N_m is total number of meso-domains and V is the total volume of the considered sample. Figure 3.17 compares the two volume-average stress tensors, $[\Sigma]^I$ and $[\Sigma]^{II}$, to the macroscopic stress tensor, Σ, defined from the boundary conditions of sample A during the biaxial loading. It shows that method II leads to a discrepancy between the average-stress tensor $[\Sigma]^{II}$ and Σ (this method overestimates the macroscopic stress by approximately 6%), while method I leads to a correct approximation of Σ. It should be noted that the discrepancy between the average stress tensor obtained by method II, $[\Sigma]^{II}$ and the macroscopic stress, Σ, results from the non-additivity of the set of volumes $\hat{V}^m$ as mentioned previously, but it does not result from the sample state. Therefore, this discrepancy would exist for any sample.

As a first approximation, method I seems to be thus more appropriate than method II to estimate Σ. However, its major drawback lies in the fact that it does not ensure the static equilibrium of meso-domains. It is worth mentioning that both methods I and II give correct estimates of the most usual variable considered in mechanics of granular materials, that is the macroscopic stress ratio, Q/P, as shown in Figure 3.18.

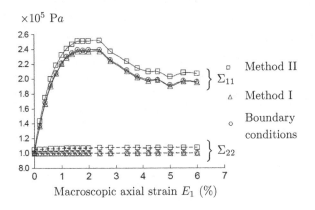

Figure 3.17. *Macroscopic stresses Σ_{11} and Σ_{22} estimated by methods I and II, compared to the respective values defined from the boundary conditions*

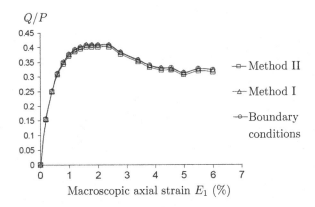

Figure 3.18. *Macroscopic stress ratio, Q/P, estimated by methods I and II, compared to the value defined from the boundary conditions*

To conclude, methods I and II are equivalent at the meso-scale and provide a correct estimate of the macroscopic stress ratio, Q/P. They are both appropriate for further analyses. However, method II exhibits a clearer physical sense; therefore, we use this method to estimate all the stress-related quantities in what follows.

3.3.4. Heterogeneity of the meso-stress

We study here the heterogeneity of the local stress field defined by method II (the results obtained for the local stress field defined by method I would be quite similar).

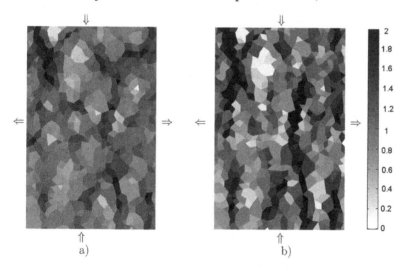

a)   b)

Figure 3.19. *Spatial distribution of the meso-stress field defined by method II for the peak state. The vertical and horizontal arrows represent the compression and extension directions respectively. The gray-level represents the value of p^m (respectively q^m) normalized by the macroscopic mean stress, P (respectively the macroscopic deviatoric stress Q)*

Forces in a granular material are known in the literature to be transmitted through the contacts in a heterogeneous manner, within a network of so called *strong* and *weak forces chains* [RAD 98, KRU 02]. We here analyze the heterogeneity at the meso-scale in terms of meso-stress distribution. Figure 3.19 shows the spatial distribution of two stress-related quantities for the peak state: the meso-mean stress, p^m, and the meso-deviatoric stress, q^m, for each meso-domain. This figure emphasizes a significantly heterogeneous meso-stress field. Moreover, the meso-stresses (the meso-deviatoric stress, in particular) are concentrated within distinct chains oriented preferentially in the compression direction, as shown for the strong forces chains by [RAD 98]. These strong stress chains thus support a major part of the external loading.

[NGU 12] have analyzed the heterogeneity of the meso-stress field by using the three dimensionless scalar measures initially proposed by [KUH 03] to quantify the heterogeneity of the meso-strain field. [NGU 12] found that the heterogeneity of the meso-stress field is more or less constant during the loading.

4

Texture–Stress–Strain Relationship at the Meso-scale

In the following, different variables characterizing structure, stress and strain of the six phases $\mathcal{W}1'$, $\mathcal{W}0'$, $\mathcal{W}2'$, $\mathcal{S}1'$, $\mathcal{S}0'$ and $\mathcal{S}2'$ are defined as the average of their counterpart over all the meso-domains belonging to each phase. The definition of porosity ϕ^p, stress tensor σ^p, and strain tensor $\delta\varepsilon^p$ for each phase p makes use of the volume-weighted average, while the definition of mean valence r^p, makes use of the arithmetic average.

4.1. Evolution of internal state

4.1.1. Evolution of meso-compactness

Figure 4.1 shows the evolution of the mean porosity of each phase, ϕ^p, during the biaxial compression test, compared to that of the porosity of the whole sample A. A high heterogeneity in density inside the sample can be observed. At the initial state, the weakly elongated phases ($\mathcal{W}1'$, $\mathcal{W}0'$ and $\mathcal{W}2'$) are looser than the sample, whereas the strongly elongated phases ($\mathcal{S}1'$, $\mathcal{S}0'$ and $\mathcal{S}2'$) are denser than the sample. In addition, the porosity of all the phases appears to be independent of their orientation at this state. However, during the deviatoric loading, the density of each phase evolves differently and depends significantly on the orientation of each phase. The phases oriented in compression direction $1'$ ($\mathcal{W}1'$ and $\mathcal{S}1'$) become much looser than the phases oriented obliquely ($\mathcal{W}0'$ and $\mathcal{S}0'$) which become, in turn, much looser than the phases oriented in extension direction $2'$ ($\mathcal{W}2'$ and $\mathcal{S}2'$). Additionally, the density of the weakly elongated phases ($\mathcal{W}1'$, $\mathcal{W}0'$, $\mathcal{W}2'$) depends more on their orientation than that of the strongly elongated phases ($\mathcal{S}1'$, $\mathcal{S}0'$ and $\mathcal{S}2'$). We can remark that phase $\mathcal{W}2'$ follows such a long compaction process that it becomes denser than the sample, while the porosity of phase $\mathcal{S}2'$ remains approximately constant during loading. Throughout the loading, the porosity of the strongly elongated phases remains lower than the porosity of the sample, whereas the porosity of the weakly elongated phases remains higher than the porosity of the sample, except for phase $\mathcal{W}2'$.

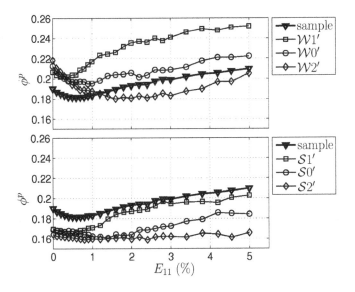

Figure 4.1. *Evolution of the porosity, ϕ^p, of the six phases $\mathcal{W}1'$, $\mathcal{W}0'$, $\mathcal{W}2'$, $\mathcal{S}1'$, $\mathcal{S}0'$ and $\mathcal{S}2'$*

Figure 4.2 shows the evolution of mean valence, r^p, of the six phases during loading. On the whole, the mean valence r^p of the strongly elongated phases ($S1'$, $S0'$, $S2'$) is higher than the mean valence, R, defined for the whole sample, whereas the mean valence, r^p, of the weakly elongated phases ($W1'$, $W0'$, $W2'$) is lower than R. Like the phase porosity, ϕ^p, the mean valence, r^p, is independent of the phase orientation at the initial state, but it becomes dependent on the orientation as soon as the loading is deviatoric. The phase $S1'$ (respectively $W1'$) has the highest mean valence r^p among the strongly elongated phases S (respectively the weakly elongated phases W), whereas the phase $S2'$ (respectively $W2'$) has the lowest mean valence r^p among the phases S (respectively W).

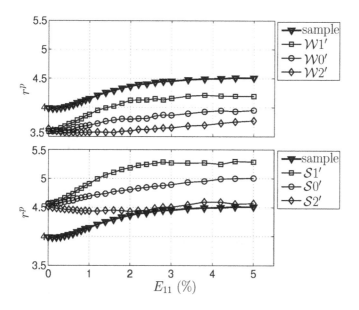

Figure 4.2. *Evolution of the valence, r^p, of the six phases $W1'$, $W0'$, $W2'$, $S1'$, $S0'$ and $S2'$*

The loading/unloading biaxial test performed on sample B (presented in section 2.2.2) allows us to study the evolution of the six phases during a loading path and an unloading path. It is worth mentioning that for the unloading path direction $1'$ changes and then the identification of the phases changes: phase $\mathcal{S}1'$ (respectively $\mathcal{W}1'$) becomes $\mathcal{S}2'$ (respectively $\mathcal{W}2'$) and phase $\mathcal{S}2'$ (respectively $\mathcal{W}2'$) becomes $\mathcal{S}1'$ (respectively $\mathcal{W}1'$).

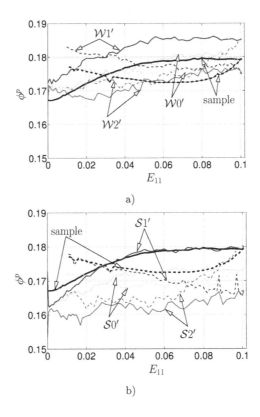

Figure 4.3. *Evolution of the porosity, ϕ^p, of the six phases during the loading path (solid line) and during the unloading path (dashed line) performed on sample B: a) for $\mathcal{W}1'$ (blue line), $\mathcal{W}0'$ (green line) and $\mathcal{W}2'$ (red line) and b) for $\mathcal{S}1'$ (blue line), $\mathcal{S}0'$ (green line) and $\mathcal{S}2'$ (red line). The porosity of sample B is represented by the black line. For a color version of this figure, see www.iste.co.uk/cambou/mesoscale.zip*

Figure 4.3 shows the evolution of the porosity, ϕ^p, of the six phases during the loading and unloading paths performed on sample B. It is shown that both samples A and B give a similar evolution of the six phases, porosity during the loading path. Phase $S1'$ oriented in the compression direction during the loading path presents an increasing porosity; however, it becomes phase $S2'$ oriented in the extension direction during the unloading path and its porosity decreases. On the other hand, phase $S2'$ oriented in the extension direction during the loading path becomes phase $S1'$ oriented in the compression direction during the unloading path and as a consequence, its porosity increases quickly. These results obtained with sample B, which are complementary to those obtained with sample A (Figure 4.1), allow us to conclude that the porosity of the six phases evolves differently during the loading and unloading paths, depending on their elongation degree and their orientation with respect to the compression direction. Whatever the loading or unloading, it is clear that the porosity of phases $W1'$ and $S1'$ oriented in the compression direction increases quickly.

4.1.2. Evolution of meso-texture

A well-known consequence of performing a deviatoric loading on a granular material whose nature is discrete is the induced anisotropy in terms of contact orientation distribution. Such a loading leads to an increase in the number of contacts oriented in the compression direction, and to a decrease in the number of contacts oriented in the extension direction (Figure 2.5). As a consequence, the distribution of meso-domain orientation changes, as shown in Figure 4.4. Each polar diagram presents the number $N(\theta)$ of meso-domains oriented in a given polar angle interval. It is clear that at the initial state, where the granular material is isotropic, meso-domains in the particle graph are isotropically oriented regardless of the meso-domain elongation degree. At the peak state, meso-domains are preferentially oriented in the compression direction, which is also the preferential direction of the distribution of contact orientation. Furthermore, this preferential distribution of meso-domain orientation is much more marked for the strongly elongated meso-domains (class $\mathcal{S}$) than for the weakly elongated meso-domains (class $\mathcal{W}$). Figure 4.6 indicates a similar distribution of meso-domain orientation in samples C and D simulated with the CD approach. This means that the preferential distribution of meso-domain orientation occurs during a deviatoric loading for any particle shape.

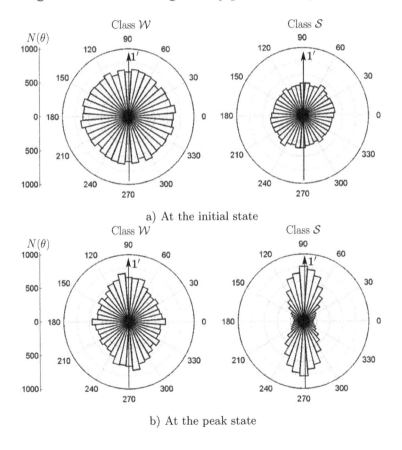

a) At the initial state

b) At the peak state

Figure 4.4. *Polar representation of the distribution of meso-domain orientation for the two elongation classes $\mathcal{W}$ and $\mathcal{S}$ a) at the initial state and b) at the peak state*

Figure 4.6 shows the distribution of meso-domain orientation during the loading and unloading test performed on sample B. We can observe an inversion of the distribution of meso-domain orientation during the unloading. In particular, the distribution of meso-domain orientation at the characteristic state during the unloading is similar to the one observed at the characteristic state during the loading, but with a rotation of $\pi/2$.

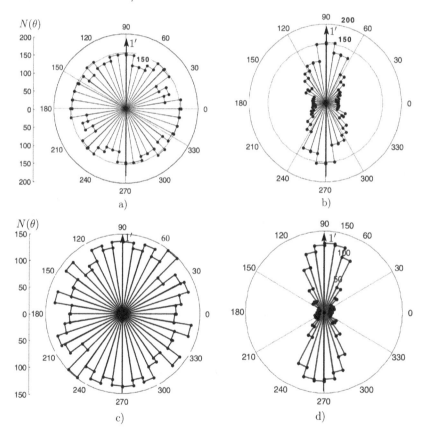

Figure 4.5. *Polar representation of the meso-domain orientation distribution for samples C (plots a and b) and D (plots c and d) at the initial state (plots a and c) and at the peak state (plots b and d)*

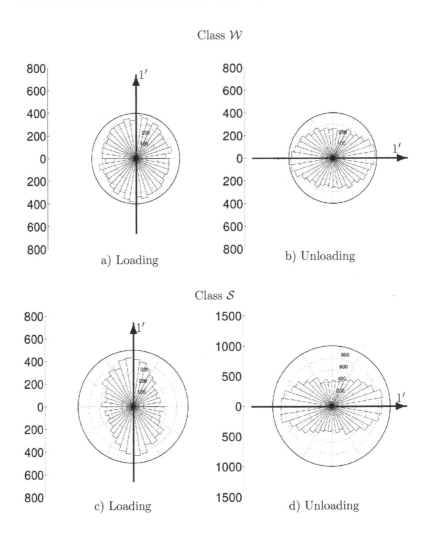

Figure 4.6. *Polar representation of the distribution of meso-domain orientations for classes $\mathcal{W}$ and $\mathcal{S}$ at the characteristic state during the loading path (a and c) and during the unloading path (b and d) performed on sample B*

Let P_v^p be the volume fraction occupied by each phase in the sample. As a consequence of the preferential distribution of meso-domain orientations observed above, the volume fraction of the six phases $W1'$, $W0'$, $W2'$, $S1'$, $S0'$ and $S2'$ changes gradually during the biaxial compression performed on sample A as shown in Figure 4.7. At the initial state where the sample is isotropic, all the phases occupy approximately equal volume fractions. As the deviatoric level of the loading increases, an increase in the volume of the phases oriented in the compression direction ($S1'$ and $W1'$) and a decrease in the volume of the phases oriented in the extension direction ($S2'$ and $W2'$) are observed, especially for the strongly elongated phases $S1'$ and $S2'$. The volume fractions, P_v^p, of the obliquely oriented phases ($S0'$ and $W0'$) remain more or less constant. This change in volume fraction of the six phases reflects the evolution of the anisotropy of the sample.

The unloading path applied on sample B leads to a strong evolution of the contact orientation distribution as shown in Figure 2.8. Meso-domains tend to be preferentially oriented in compression direction $1'$ (Figure 4.6). As shown in Figure 4.8, phases $W1'$ and $S1'$ during the loading become phases $W2'$ and $S2'$ during the unloading and their volume fraction P_v^p decreases. On the other hand, phases $W2'$ and $S2'$ become $W1'$ and $S1'$ and their volume fraction increases. It is interesting to note that the volume fraction of phases $W0'$ and $S0'$ remains more or less constant during the loading and unloading. Figure 4.8 also points out that the values of P_v^p at the critical state during the loading and the unloading are almost the same for each phase. For example, the value of P_v^p for phase $S1'$ is 0.33 at the critical state for the loading and 0.34 at the same state for the unloading.

The contribution of each phase to the anisotropy of the sample in terms of the distribution of meso-domain orientation can be quantified by the following tensor, called the *texture tensor*:

$$\boldsymbol{T}^p = \frac{1}{V} \sum_{m \in \mathcal{C}^p} V^m \boldsymbol{m}^m \otimes \boldsymbol{m}^m, \qquad\qquad [4.1]$$

where the superscript m runs over the set C^p of meso-domains belonging to phase p. The texture tensor of the sample can be obtained by summing its counterparts defined for each of the six phases $\mathcal{W}1'$, $\mathcal{W}0'$, $\mathcal{W}2'$, $\mathcal{S}1'$, $\mathcal{S}0'$ and $\mathcal{S}2'$:

$$T = \sum_{p=1}^{6} T^p = \frac{1}{V} \sum_{m=1}^{N_m} V^m m^m \otimes m^m. \qquad [4.2]$$

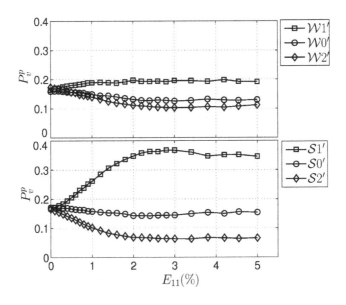

Figure 4.7. *Evolution of the volume fraction, P_v^p, of the six phases $\mathcal{W}1'$, $\mathcal{W}0'$, $\mathcal{W}2'$, $\mathcal{S}1'$, $\mathcal{S}0'$ and $\mathcal{S}2'$*

It is observed that, during a biaxial loading or unloading, the off-diagonal components of T^p and T are very small compared to the diagonal ones. In this case, the anisotropy level of the sample and of each phase can be quantified by the respective deviatoric textures, T_d and T_d^p, defined for the sample and for each phase:

$$T_d^p = T_{11}^p - T_{22}^p \quad \text{and} \quad T_d = T_{11} - T_{22}. \qquad [4.3]$$

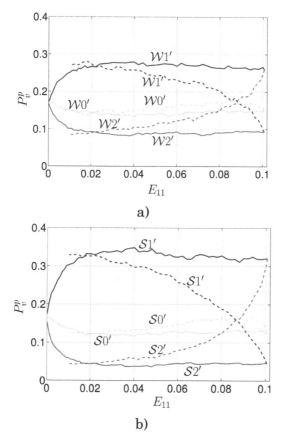

Figure 4.8. *Evolution of the volume fraction, P_v^p, of the six phases $\mathcal{W}1'$, $\mathcal{W}0'$, $\mathcal{W}2'$, $\mathcal{S}1'$, $\mathcal{S}0'$ and $\mathcal{S}2'$ during a) the loading and b) the unloading paths applied on sample B. Blue lines: phases $\mathcal{W}1'$ and $\mathcal{S}1'$; red lines: phases $\mathcal{W}2'$ and $\mathcal{S}2'$; green lines: phases $\mathcal{W}0'$ and $\mathcal{S}0'$; solid lines: loading; dashed lines: unloading. For a color version of this figure, see www.iste.co.uk/cambou/mesoscale.zip*

It is easy to obtain:

$$T_d = \sum_{p=1}^{6} T_d^p \qquad [4.4]$$

The deviatoric texture, T_d^p, can be thought of as being the contribution of each phase to the anisotropy of the sample. Figure 4.9 shows the evolution of the contribution, T_d^p, of each phase to the anisotropy of the sample. The deviatoric texture, T_d, for the whole sample (dash-dotted line) is divided by six as if the six phases contributed equally to the anisotropy of the sample. The increase in the deviatoric texture, T_d, for the whole sample confirms that meso-domains become preferentially oriented in the compression direction. It is clear that phases $S1'$ and $W1'$ contribute positively to the anisotropy, while phases $S2'$ and $W2'$ contribute negatively; and phases $S0'$ and $W0'$ do not induce any anisotropy to the sample. Additionally, the contribution of phase $S1'$ increases quickly and becomes dominant at a large strain. Figure 4.10 indicates similar results for sample D composed of octagonal particles.

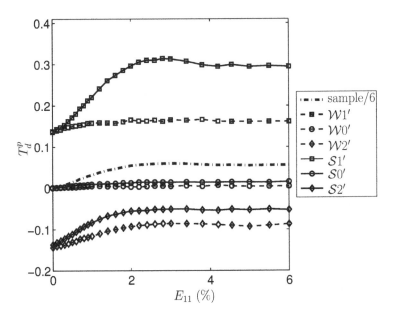

Figure 4.9. *Evolution of the deviatoric texture, T_d^p, of the six phases $W1'$, $W0'$, $W2'$, $S1'$, $S0'$ and $S2'$, compared to the deviatoric texture T_d of sample A*

The dominant contribution of phases $W1'$ and $S1'$ is also observed during the loading and unloading paths applied on sample B as shown in Figure 4.11. It can be noted in

Figure 4.11 that the value of T_d^p for phase $\mathcal{S}1'$ (respectively $\mathcal{S}2'$) at the critical state during loading is opposite to the value of T_d^p for phase $\mathcal{S}1'$ (respectively $\mathcal{S}2'$) at the same state during the unloading. This is due to the fact that T_d and T_d^p have been defined in the fixed axes 1 and 2 to produce more understandable figures.

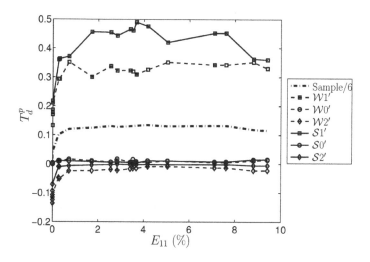

Figure 4.10. *Evolution of the deviatoric texture, T_d^p, of the six phases $\mathcal{W}1'$, $\mathcal{W}0'$, $\mathcal{W}2'$, $\mathcal{S}1'$, $\mathcal{S}0'$ and $\mathcal{S}2'$, compared to the deviatoric texture, T_d, for the whole sample D*

4.2. Strain–texture relationship

The relationship between strain and texture for granular materials at the macro-scale is well-known in the literature. An isotropic assembly deforms differently from an anisotropic one [ODA 72, KON 82]. Moreover, for a given anisotropic assembly, the deformation depends strongly on the loading direction [CAM 88b]. In this section, we investigate whether there is a similar relation between strain and texture at the meso-scale.

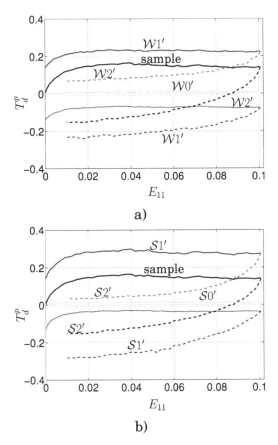

Figure 4.11. *Deviatoric texture, T_d^p, of the six phases $\mathcal{W}1'$, $\mathcal{W}0'$, $\mathcal{W}2'$, $\mathcal{S}1'$, $\mathcal{S}0'$ and $\mathcal{S}2'$, compared to the deviatoric texture, T_d, of sample B and b) the loading and b) the unloading paths. For comparison, one sixth of T_d is plotted. Deviatoric textures, T_d and T_d^p, are defined in the fixed axes 1 and 2. Blue lines: phases $\mathcal{W}1'$ and $\mathcal{S}1'$; red lines: phases $\mathcal{W}2'$ and $\mathcal{S}2'$; green lines: phases $\mathcal{W}0'$ and $\mathcal{S}0'$; solid lines: loading; dashed-lines: unloading. For a color version of this figure, see www.iste.co.uk/cambou/mesoscale.zip*

To analyze the relation between texture and strain at the meso-scale, a volume-weighted average strain tensor is calculated over a set of all meso-domains with similar textures (the six phases). This average operation works as long as the strain is not strongly localized in shear bands.

Otherwise, we need a special average operation performed only on the meso-domains within shear bands; however, this goes beyond the scope of the analysis presented in this section. As shown in section 3.2, strain is strongly localized in shear bands after the peak of the stress-strain curve. Therefore, we perform our analysis only for the loading zone before the peak of the stress-strain curve (before the axial strain $E_{11} = 2\%$ of sample A).

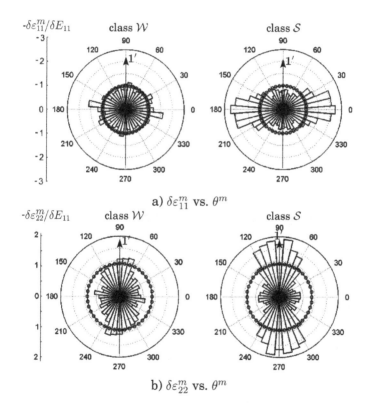

a) $\delta\varepsilon_{11}^{m}$ vs. θ^{m}

b) $\delta\varepsilon_{22}^{m}$ vs. θ^{m}

Figure 4.12. *Opposite of the increments of strain components $\delta\varepsilon_{11}^{m}$ and $\delta\varepsilon_{22}^{m}$ normalized by the prescribed axial strain increment, δE_{11}, versus the meso-domain orientation, θ^{m}, for the two meso-domain classes $\mathcal{W}$ and $\mathcal{S}$ for the strain increment prescribed at $E_{11} = 1.2\%$: a) for $\delta\varepsilon_{11}^{m}$ and b) for $\delta\varepsilon_{22}^{m}$*

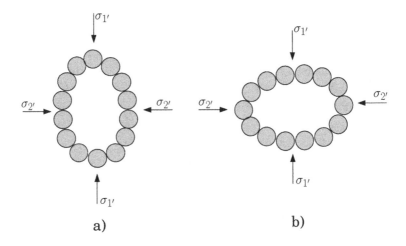

Figure 4.13. *During a compression test in direction $1'$, meso-domains a) oriented in direction $1'$ give rise to a greater value of $|\delta\varepsilon_{22}^m|$ than meso-domains b) oriented in direction $2'$. On the contrary, meso-domains b) give rise to a greater value of $|\delta\varepsilon_{11}^m|$ than meso-domains a)*

The polar diagrams in Figure 4.12 represent the increment of strain components, $\delta\varepsilon_{11}^m$ and $\delta\varepsilon_{22}^m$, versus the meso-domain orientation, θ^m, for the two meso-domain elongation classes $\mathcal{W}$ and $\mathcal{S}$. For this polar representation, all the meso-domains in each class are split, according to the value of θ^m, into 40 sets corresponding to 40 orientation angle intervals, and a volume-weighted average strain tensor is calculated for each set. Note that $\delta\varepsilon_{11}^m$ and $\delta\varepsilon_{22}^m$ are compression and extension strain increments, respectively. The increment of shear strain component, $\delta\varepsilon_{12}^m$, is negligible compared to $\delta\varepsilon_{11}^m$ and $\delta\varepsilon_{22}^m$, meaning that the principal directions of strain at the meso-scale coincide more or less with those of strain at the macro-scale. Figure 4.12 shows that the increment of meso-strains $\delta\varepsilon_{11}^m$ and $\delta\varepsilon_{22}^m$ depend significantly on the meso-domain orientation, especially for class $\mathcal{S}$ of strongly elongated meso-domains. The increment of the compression strain $\delta\varepsilon_{11}^m$ is larger within the meso-domains oriented in the extension direction than within those oriented in the compression direction. In contrast, the increment of the extension strain $\delta\varepsilon_{22}^{m}$ is larger within the meso-domains oriented in the compression direction than within those oriented in the extension direction. These phenomena are illustrated in Figure 4.13.

Figure 4.14 shows the increment of the volumetric strain, $\delta\varepsilon_{v}^{m}$, versus the meso-domain orientation, θ^{m}, for the two meso-domain elongation classes $\mathcal{W}$ and $\mathcal{S}$. Two loading increments are prescribed at $E_{11} = 0.2\%$ and at $E_{11} = 1.2\%$ at which the sample contracts and dilates, respectively. It is shown that all meso-domains contract at the first loading increment and those oriented in the extension direction contract more than those oriented in the compression direction. At the second loading increment, meso-domains oriented in the compression direction show a dilative behavior, while those oriented in the extension direction show a contractive behavior. This result is in good agreement with the result reached by [KUH 99]. Furthermore, the volumetric behavior of the strongly elongated meso-domains is much more marked than that of the weakly elongated meso-domains.

Figures 4.15 and 4.16 show the respective ratios of the strain components $\delta\varepsilon_{11}^{p}$ and $\delta\varepsilon_{22}^{p}$ for each phase to their macroscopic counterparts δE_{11} and δE_{22} versus the macroscopic axial strain E_{11}. Values of $\delta\varepsilon_{11}^{p}/\delta E_{11}$ for phases $\mathcal{W}2'$, $\mathcal{S}2'$ are higher than 1.0, while those for phases $\mathcal{W}0'$, $\mathcal{S}0'$, $\mathcal{W}1'$ and $\mathcal{S}1'$ are lower than 1.0 throughout loading. Additionally, the ratio $\delta\varepsilon_{11}^{p}/\delta E_{11}$ for phase $\mathcal{S}2'$ is much higher than for the other phases and increases greatly during loading. On the other hand, the ratio $\delta\varepsilon_{22}^{p}/\delta E_{22}$ for phases $\mathcal{W}1'$ and $\mathcal{S}1'$ is higher than 1.0, while the one for phases $\mathcal{W}0'$, $\mathcal{S}0'$, $\mathcal{W}2'$ and $\mathcal{S}2'$ is lower than 1.0. In particular, the ratio $\delta\varepsilon_{22}^{p}/\delta E_{22}$ for phase $\mathcal{S}1'$ increases to a large extent, while the one for phase $\mathcal{S}2'$ decreases during loading. These results confirm that the phases oriented in the compression direction ($\mathcal{W}1'$ and $\mathcal{S}1'$) deform more in the extension direction, while those oriented in the extension direction ($\mathcal{W}2'$ and $\mathcal{S}2'$) deform more in the compression direction. The phases oriented obliquely ($\mathcal{W}0'$ and $\mathcal{S}0'$) more or less follow the deformation of the sample.

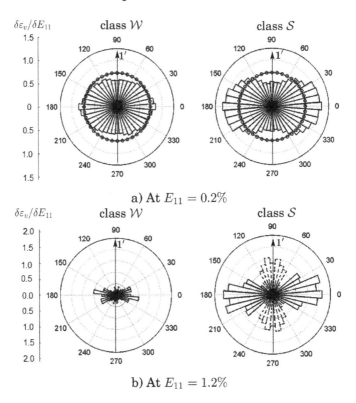

a) At $E_{11} = 0.2\%$

b) At $E_{11} = 1.2\%$

Figure 4.14. *Increments of volumetric strain, $\delta\varepsilon_v^m$, normalized by the prescribed axial strain increment, δE_{11}, versus the meso-domain orientation, θ^m, for the two meso-domain classes $\mathcal{W}$ and $\mathcal{S}$ for two strain increments prescribed at a) $E_{11} = 0.2\%$ and b) $E_{11} = 1.2\%$. The solid line and the dashed line depict the contraction and dilation, respectively. The circles with marker (○) depict the macroscopic volumetric strain*

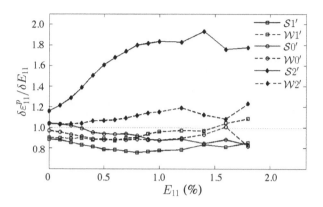

Figure 4.15. *The ratio $\delta\varepsilon^p_{11}/\delta E_{11}$ versus the macroscopic axial strain, E_{11}, for the six phases $\mathcal{W}1'$, $\mathcal{W}0'$, $\mathcal{W}2'$, $\mathcal{S}1'$, $\mathcal{S}0'$ and $\mathcal{S}2'$*

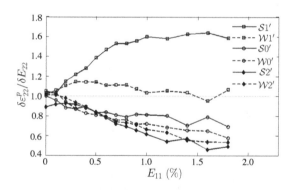

Figure 4.16. *The ratio $\delta\varepsilon^p_{22}/\delta E_{22}$ versus the macroscopic axial strain, E_{11}, for the six phases $\mathcal{W}1'$, $\mathcal{W}0'$, $\mathcal{W}2'$, $\mathcal{S}1'$, $\mathcal{S}0'$ and $\mathcal{S}2'$*

The volumetric behavior of each phase can be characterized by a scalar variable r^p_ε called the *phase strain ratio*:

$$r^p_\varepsilon = \frac{\delta\varepsilon^p_v}{\delta\varepsilon^p_d} = \frac{\delta\varepsilon^p_{11} + \delta\varepsilon^p_{22}}{\mid \delta\varepsilon^p_{11} - \delta\varepsilon^p_{22} \mid}. \qquad [4.5]$$

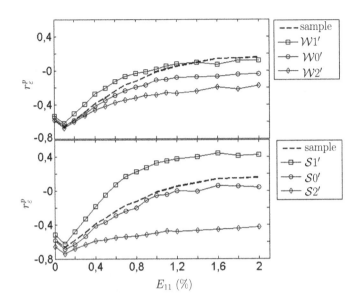

Figure 4.17. *The phase strain ratio, r_ε^p, versus the macroscopic axial strain, E_{11}, for the six phases $\mathcal{W}1'$, $\mathcal{W}0'$, $\mathcal{W}2'$, $\mathcal{S}1'$, $\mathcal{S}0'$ and $\mathcal{S}2'$*

A negative and a positive value of r_ε^p means contraction and dilation of the phase under consideration, respectively. Figure 4.17 shows the evolution of the phase strain ratio, r_ε^p, of the six phases during loading, compared to the strain ratio $\delta E_v/\delta E_d$ defined for the sample. We can see that the local strain within the six phases evolves differently from the macroscopic strain in a way which is significantly dependent on the phase orientation, especially for the strongly elongated phases ($\mathcal{S}1'$, $\mathcal{S}0'$ and $\mathcal{S}2'$). Phase $\mathcal{S}1'$ dilates greatly like a dense material, whereas phase $\mathcal{S}2'$ contracts throughout the loading like a loose material. Phase $\mathcal{S}0'$ behaves like a material with an intermediate density; moreover, its volumetric behavior is quite close to that of the sample. This observation also holds during the unloading path applied on sample B as shown in Figure 4.18. It should be noted that the curves in Figure 4.18 are quite fluctuating, compared to those in Figure 4.17. This might be explained as follows. Firstly, the high contact stiffness used to simulate sample B greatly reduces the contact duration between two particles concerned, leading to a significant fluctuation of the local incremental displacement field. Secondly, the strain ratio r_ε^p for each phase is calculated up to $E_{11} = 10\%$ (after the peak of the stress-strain curve in Figure 2.7). The local strain is localized in shear bands after the peak state, which causes the strain of each phase to fluctuate. When the CD approach is used, the analysis of local deformation at the meso-scale becomes problematic as the particle displacement field fluctuates strongly during the simulation. One can see in Figure 4.19 that increments of the volumetric strain $\delta\varepsilon_v^p$ defined for phases $\mathcal{W}1'$ and $\mathcal{S}1'$ and δE_v defined for sample D fluctuate greatly during the simulation. For the CD approach, particles are assumed to be perfectly rigid, leading to a jump in velocity when two particles collide each other and then to a strong fluctuation of the local displacement field. Fluctuation of the local strain field notwithstanding, one can observe that the dilatancy of phase $\mathcal{S}1'$ is significantly higher than that of phase $\mathcal{W}1'$ and of sample D.

4.3. Stress–texture relationship

As shown in Figure 3.19, stress is significantly heterogeneous at the meso-scale. It is found that the stress state at the meso-scale is anisotropic even when the macroscopic stress state is isotropic at the initial state. It can be seen in Figure 4.20(a) that the local stress ratio, q^m/p^m, defined for each meso-domain is, significantly different from zero at this state, especially for the strongly elongated meso-domains for which $q^m/p^m \approx 0.14$. Furthermore, the local stress ratio, q^m/p^m, is independent of the meso-domain orientation, θ^m, at the initial state as shown in Figure 4.20(a). Concerning the local principal stress directions, they are governed by the meso-domain orientation. In Figure 4.20(b), the orientation of the local major principal stress direction, $n_{\sigma_1}^m$, is plotted versus the meso-domain orientation, θ^m. It is shown that the plotted points are localized more or less on the diagonal for the macroscopic stress ratio $Q/P = 0$, indicating that the orientation of the major principal stress direction, $n^m_{\sigma_1}$, at the meso-scale coincides with the meso-domain orientation, θ^m, at the initial state. Furthermore, as the level of the deviatoric loading increases, the local major principal stress direction tends to rotate toward the major principal stress direction at the macro-scale ($\theta^m = 0^o$). However, a little discrepancy between the principal stress directions at the meso-scale and at the macro-scale is still observed even at the peak state ($Q/P = 0.41$). It is noted that, unlike stress at the meso-scale, the principal strain directions at this scale are mainly governed by those at the macro-scale, even at the initial state as mentioned in section 4.2.

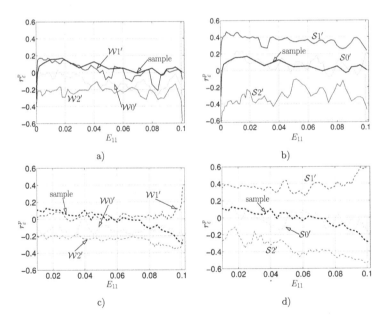

Figure 4.18. *The phase strain ratio, r^p_ε, versus the macroscopic axial strain, E_{11}, for the six phases $W1'$, $W0'$, $W2'$, $S1'$, $S0'$ and $S2'$ during (a, b) the loading and (c, d) the unloading applied on sample B. Blue lines: phases $W1'$ and $S1'$; red lines: phases $W2'$ and $S2'$; green lines: phases $W0'$ and $S0'$; solid lines: loading; dashed-lines: unloading. For a color version of this figure, see www.iste.co.uk/cambou/mesoscale.zip*

Figure 4.21 shows the local mean stress, p^m, and the local stress ratio, q^m/p^m, versus the meso-domain orientation, θ^m, for the two elongation classes, $\mathcal{W}$ and $\mathcal{S}$, at the peak state. The corresponding macroscopic values are represented by circles. The figure shows that the local mean stress, p^m, is independent of the meso-domain elongation degree and orientation; and it is quite close to the macroscopic value, P. On the other hand, the local stress ratio, q^m/p^m, which quantifies the level of the deviatoric stress at the meso-scale, depends heavily on the meso-domain orientation, θ^m, especially for the strongly elongated meso-domains (class $\mathcal{S}$). The meso-domains oriented in the compression direction bear a higher deviatoric stress than those oriented in the extension direction. In addition, the stress ratio, q^m/p^m, within the former ones is higher than the macroscopic value, Q/P, while q^m/p^m within the latter ones is smaller than Q/P. These results mean that the mean stress and the deviatoric stress applied to a granular assembly are transmitted differently through meso-domains. The transmission of the mean stress is independent of the meso-domain texture, while the transmission of the deviatoric stress is greatly dependent on the meso-domain texture.

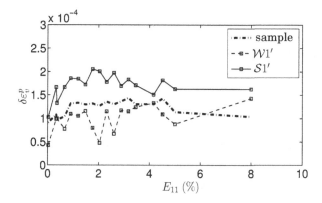

Figure 4.19. *Increments of the volumetric strain, $\delta\varepsilon_v^p$, for phases $\mathcal{W}1'$ and $\mathcal{S}1'$ versus the macroscopic axial strain, E_{11}, compared to the increment of the macroscopic volumetric strain, δE_v, for sample D*

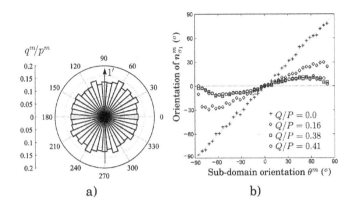

a)   b)

Figure 4.20. *a) The stress ratio, q^m/p^m, versus the meso-domain orientation, θ^m, for class $\mathcal{S}$ of strongly elongated meso-domains at the initial state and b) the orientation of the local major principal stress direction, $\mathbf{n}_{\sigma_1}^m$, versus the meso-domain orientation, θ^m, for different values of the macroscopic stress ratio, Q/P*

The local stress ratio, q^p/p^p, defined for each phase does not fully give information about how each phase participates in supporting the deviatoric stress at the macro-scale. The global stress ratio, Q/P, cannot be derived from the local one by a volume-weighted average.

Let us develop the relation between the global stress ratio, Q/P and local stresses:

$$Q/P = \frac{\Sigma_{11} - \Sigma_{22}}{\Sigma_{11} + \Sigma_{22}} = \frac{\sum_{p=1}^{6} V^p(\sigma_{11}^p - \sigma_{22}^p)}{\sum_{p=1}^{6} V^p(\sigma_{11}^p + \sigma_{22}^p)}. \qquad [4.6]$$

It was shown earlier that the local mean stress is independent of the meso-domain texture, that is to say that the term $\sigma_{11}^p + \sigma_{22}^p$ in the denominator of equation [4.6] is a constant. As a consequence, equation [4.6] can be rewritten as:

$$Q/P = \frac{1}{V}\sum_{p=1}^{6} V^p \frac{\sigma_{11}^p - \sigma_{22}^p}{\sigma_{11}^p + \sigma_{22}^p} = \frac{1}{V}\sum_{p=1}^{6} V^p r_\sigma^p, \qquad [4.7]$$

where r_σ^p is defined for each phase as follows:

$$r_\sigma^p = \frac{\sigma_{11}^p - \sigma_{22}^p}{\sigma_{11}^p + \sigma_{22}^p}. \qquad [4.8]$$

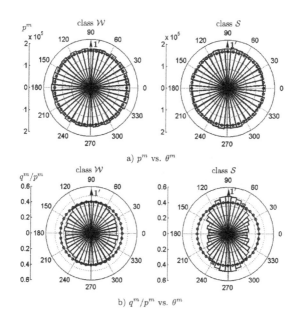

a) p^m vs. θ^m

b) q^m/p^m vs. θ^m

Figure 4.21. a) Mean stress, p and b) stress ratio, q^m/p^m, versus the meso-domain orientation, θ^m, for the two meso-domain classes $\mathcal{W}$ and $\mathcal{S}$ at the peak state

It should be noted that, for a better understanding, the variable r_σ^p is defined in the fixed axes 1 and 2. The new introduced variable, r_σ^p, is called the *phase stress ratio* and is complementary to the local stress ratio, q^p/p^p and can be interpreted as the participation of each phase into the global stress ratio, Q/P. When the principal stress directions at the meso-scale coincide with those at the macro-scale, the phase stress ratio, r_σ^p, is equal to the local stress ratio, q^p/p^p. As shown previously, this is the case for a high level of the deviatoric loading at the macro-scale.

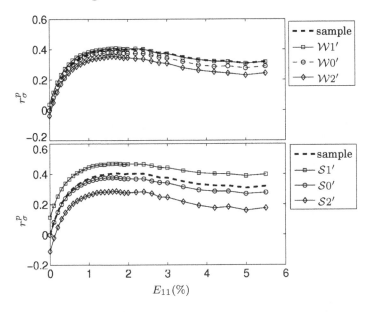

Figure 4.22. *The phase stress ratio, r_σ^p, versus the macroscopic axial strain, E_{11}, for the six phases $\mathcal{W}1'$, $\mathcal{W}0'$, $\mathcal{W}2'$, $\mathcal{S}1'$, $\mathcal{S}0'$ and $\mathcal{S}2'$ of sample A*

The evolution of the phase stress ratio, r_σ^p, defined for the six phases $\mathcal{W}1'$, $\mathcal{W}0'$, $\mathcal{W}2'$, $\mathcal{S}1'$, $\mathcal{S}0'$ and $\mathcal{S}2'$ during the biaxial compression test performed on sample A is shown in Figure 4.22. It can be seen that, at the initial state where the macroscopic stress ratio Q/P is equal to zero, values of r_σ^p for the six phases are very different: positive values for phases $\mathcal{W}1'$ and $\mathcal{S}1'$, negative values for phases $\mathcal{W}2'$ and $\mathcal{S}2'$ and almost zero for phases $\mathcal{W}0'$ and $\mathcal{S}0'$. Negative values of r_σ^p for phases $\mathcal{W}2'$ and $\mathcal{S}2'$ are due to the orientation of these phases in the extension direction, $2'$, leading to a higher value of the stress component σ_{22} than the value of the stress component σ_{11} for these phases. During the biaxial compression, the increase in the phase stress ratio, r_σ^p, of the six phases depends strongly on the phase orientation, in particular for the strongly elongated phases. The value of r_σ^p for phase $S1'$ is always higher than the macroscopic stress ratio, Q/P. The value of r_σ^p for phase $S2'$ becomes positive when the deviatoric loading is sufficiently high; it, however, remains lower than Q/P. Phase $S0'$ has a value of r_σ^p intermediate between those of phases $S1'$ and $S2'$ and close to the macroscopic value, Q/P.

The evolution of the stress ratio, r_σ^p, during the loading stage applied on sample B is similar to that obtained with sample A as shown in Figure 4.23. During the unloading stage, r_σ^p for each phase decreases and r_σ^p for the phases oriented in the compression direction ($W1'$ and $S1'$) decreases quicker than for the phases oriented in the extension direction ($W2'$ and $S2'$). It can be pointed out that the values of r_σ^p for each phase at the critical state for the unloading are approximately opposite to those at the same state for the loading.

The above results in Figures 4.22 and 4.23 emphasize the fact that when a granular material is subjected to deviatoric loading, the local phases contribute differently to support the deviatoric loading, depending on their orientation, particularly for the strongly elongated phases. Phase $S1'$ oriented in the compression direction supports much better the applied deviatoric loading than phase $S2'$ oriented in the extension direction, while the contribution of phase $S0'$ oriented obliquely is intermediate between the phases $S1'$ and $S2'$.

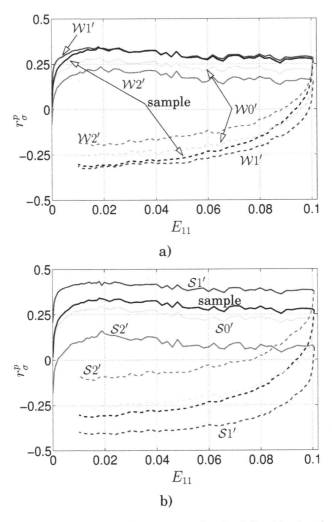

Figure 4.23. *Evolution of the phase stress ratio, r_σ^p, defined for the six phases $\mathcal{W}1'$, $\mathcal{W}0'$, $\mathcal{W}2'$, $\mathcal{S}1'$, $\mathcal{S}0'$ and $\mathcal{S}2'$ during a) the loading stage and b) the unloading stage performed on sample B. Blue lines: phases $\mathcal{W}1'$ and $\mathcal{S}1'$; red lines: phases $\mathcal{W}2'$ and $\mathcal{S}2'$; green lines: phases $\mathcal{W}0'$ and $\mathcal{S}0'$; solid lines: loading; dashed lines: unloading. For a color version of this figure, see www.iste.co.uk/cambou/mesoscale.zip*

4.4. Mechanical behavior of the granular material at the meso-scale

We will analyze the mechanical behavior of the six phases $\mathcal{S}1'$, $\mathcal{S}0'$, $\mathcal{S}2'$, $\mathcal{W}1'$, $\mathcal{W}0'$ and $\mathcal{W}2'$ in this section. The analysis of the behavior of each phase will be focused on some important states such as the initial, characteristic, peak and critical states during the compression test performed on sample A. Furthermore, the initial and critical states during the loading-unloading cycle performed on sample B are also analyzed. The stress state is studied through the phase stress ratio, r_σ^p, defined by equation [4.8] and strain state is studied through the phase strain ratio, r_ε^p, defined by equation [4.5].

	Sample	$\mathcal{W}1'$	$\mathcal{W}0'$	$\mathcal{W}2'$	$\mathcal{S}1'$	$\mathcal{S}0'$	$\mathcal{S}2'$
P_v^p		0.16	0.16	0.17	0.17	0.17	0.17
T_d^p	0.0	0.14	0.0	-0.14	0.14	0.0	-0.14
ϕ^p	0.19	0.21	0.21	0.22	0.17	0.17	0.16
r^p	4.0	3.6	3.6	3.6	4.6	4.6	4.5
r_σ^p	0.0	0.04	0.0	-0.04	0.11	0.0	-0.11
r_ε^p	-0.57	-0.53	-0.55	-0.58	-0.52	-0.59	-0.66

Table 4.1. *Values of different characteristics of sample A and of the six phases at the initial state*

4.4.1. Initial state

Table 4.1 presents some characteristics of sample A and of the six phases at the initial state, such as the volume fraction, P_v^p, the measure of anisotropy, T_d^p, the porosity, ϕ^p, valence, r^p, the stress ratio, r_σ^p and the strain ratio, r_ε^p. Note that sample A is isotropic and the global stress is also isotropic at the initial state. The isotropy of the sample is reflected by the fact that the volume of the sample is uniformly partitioned into the six phases, whatever the orientation of the phases, and that the density of the phases (porosity, ϕ^p and valence, r^p) is independent of the phase orientation. Moreover, the positive contributions of phases $\mathcal{S}1'$ and $\mathcal{W}1'$ to the global anisotropy are completely counterbalanced by the negative contributions of phases $\mathcal{S}2'$ and $\mathcal{W}2'$. The initial density of each phase depends only on the phase elongation degree: the strongly elongated phases ($\mathcal{S}1'$, $\mathcal{S}0'$ and $\mathcal{S}2'$) are denser than the weakly elongated phases ($\mathcal{W}1'$, $\mathcal{W}0'$ and $\mathcal{W}2'$), i.e. the former have a lower porosity, ϕ^p and a higher valence, r^p, than the latter.

The local stress is not uniform within the phases at the initial state, in particular within the strongly elongated phases, which is demonstrated by the dependence of the local stress ratio, r_σ^p, on the phase orientation. The global isotropic stress state is reflected by the fact that the positive values of r_σ^p for phases $\mathcal{S}1'$ and $\mathcal{W}1'$ are opposite to the negative values for phases $\mathcal{S}2'$ and $\mathcal{W}2'$. The negative values of the phase strain ratio, r_ε^p, show that all the phases are contracting at the initial state. Phase $\mathcal{S}2'$ oriented in the extension direction contracts much more than phase $\mathcal{S}1'$ oriented in the compression direction ($r_\varepsilon^p = -0.66$ for phase $\mathcal{S}2'$ compared to $r_\varepsilon^p = -0.52$ for phase $\mathcal{S}1$), in particular. It can be noted that the highly contractant behavior observed for sample A at the beginning of the loading is due to the elastic strain at contacts between particles.

	Sample	$\mathcal{W}1'$	$\mathcal{W}0'$	$\mathcal{W}2'$	$\mathcal{S}1'$	$\mathcal{S}0'$	$\mathcal{S}2'$
	Loading stage						
P_v^p		0.17	0.17	0.17	0.17	0.16	0.15
T_d^p	0.0	0.14	0.0	-0.14	0.14	0.0	-0.15
r_σ^p	0.0	0.09	0.02	-0.08	0.2	0.02	-0.18
	Unloading stage						
P_v^p		0.09	0.14	0.26	0.04	0.13	0.32
T_d	0.14	-0.08	0.02	0.23	-0.03	0.01	0.28
r_σ^p	0.28	0.16	0.22	0.3	0.08	0.22	0.38

Table 4.2. *Different characteristics of sample B and of the six phases at the initial state for the loading/unloading cycle performed on sample B*

Table 4.2 presents some characteristics of the six phases at the initial state for the loading/unloading test performed on sample B. We can see that values of the volume fraction, P_v^p, of the deviatoric texture, T_d^p and of the phase stress ratio, r_σ^p, at the initial state of the loading are very different from those at the initial state of the unloading. This is due to the anisotropy induced by the deviatoric loading path.

4.4.2. Characteristic state

The characteristic state of a phase is defined as the state at which the phase under consideration changes from contractant to dilatant behavior (thus, the phase strain ratio, r_ε^p, is zero at this state). Table 4.3 encapsulates some characteristics of the six phases and of sample A at their respective characteristic states. We can see in Table 4.3 that the six phases reach the characteristic state differently from the sample and from each other. The phases oriented in the compression direction ($S1'$ and $W1'$) reach this state before the sample, whereas the phases oriented in the extension direction ($S2'$ and $W2'$) have not yet reached the characteristic state in the studied range of loading. One can remark that the porosity, ϕ^p and the valence, r^p, of each phase at the characteristic state is practically independent of the phase orientation. Furthermore, the values of r_σ^p at the characteristic state for phases $S1'$, $W1'$, $S2'$ and $W2'$ are quite close to each other; and they are also close to the global stress ratio of the sample at this state ($Q/P \approx 0.38$). The phase stress ratio, r_σ^p, of phases $S2'$ and $W2'$, which have not yet reached the characteristic state, is lower than the global stress ratio, Q/P.

	Sample	$W1'$	$W0'$	$W2'$	$S1'$	$S0'$	$S2'$
E_{11}	1.1 %	1.0 %	2.0 %	–	0.5 %	1.2 %	–
P_v^p		0.15	0.11	–	0.21	0.16	–
T_d^p	0.19	0.16	0.0	–	0.17	0.01	–
ϕ^p	0.18	0.22	0.21	–	0.17	0.16	–
r^p	4.1	3.9	3.8	–	4.7	4.7	–
r_σ^p	0.38	3.8	0.38	–	0.37	0.37	–

Table 4.3. *Different characteristics of the six phases and of sample A at the characteristic state*

4.4.3. Peak state

The peak state of a phase is the state at which the stress ratio, r_σ^p, within the considered phase is maximum. Table 4.4

presents some characteristics of the six phases and of sample A at the peak state. We can see in Table 4.4 that all the six phases reach the peak state at approximately the same axial macroscopic strain, $E_{11} = 2\%$, at which the sample also reaches the peak state. At this state, the volume of phase $S1'$ reaches approximately the maximum value, while the volume of phase $S2'$ reaches the minimum value (Figure 4.7). As a consequence, phase $S1'$ contributes mainly to the global anisotropy ($T_d^{S1'} = 0.33$), while the role of phase $S2'$ in reducing the anisotropy created by phase $S1'$ is not important ($T_d^{S2'} = -0.06$). Moreover, the difference in the density between the phases is great at the peak state: phase $S1'$ (respectively $W1'$) is much looser than phase $S2'$ (respectively $W2'$).

	Sample	$W1'$	$W0'$	$W2'$	$S1'$	$S0'$	$S2'$
E_{11}	2.0 %	2.0 %	2.0 %	2.0 %	2.0 %	2.0 %	2.0 %
P_v^p		0.20	0.13	0.11	0.35	0.14	0.07
T_d^p	0.33	0.17	0.0	-0.09	0.30	0.01	-0.06
ϕ^p	0.19	0.24	0.21	0.18	0.19	0.17	0.16
r^p	4.4	4.1	3.8	3.6	5.3	4.8	4.4
r_σ^p	0.40	0.40	0.38	0.35	0.47	0.37	0.28
r_ε^p	0.15	0.12	-0.04	-0.18	0.42	0.04	-0.43

Table 4.4. *Values of different characteristics of sample A and of the six phases at the peak state*

We can see that the phase stress ratio, r_σ^p, of each phase is highly governed by the phase orientation at the peak state, in particular for the strongly elongated phases: r_σ^p of phase $S1'$ is much higher than r_σ^p of phase $S2'$. Another point is that the strain ratio, r_ε^p, of each phase depends highly on the phase orientation at this state. Phase $S1'$ presents a high dilatancy ($r_\varepsilon^p = 0.42$ compared to a value of 0.15 for the sample), whereas phase $S2'$ is still contracting considerably ($r_\varepsilon^p = -0.43$). In fact, the dilatancy of a granular material has been shown to be maximum at the peak state in many studies, in particular in [CAS 40, TAY 48]. In the case of the six local phases considered in the current study, we can observe that the strain ratio, r_ε^p, of each phase remains approximately constant near the peak state and seems to reach the maximum value at this state. However, this observation should be confirmed by more detailed analyses, in which the evolution of the local strain after the peak state can be studied with a suitable method which takes into account the formation of shear bands inside the sample.

4.4.4. Critical state

The critical state is the final state of a granular material under deviatoric loading, where the material evolves without any volume change. Moreover, the stress ratio remains constant at this state. It can be seen in Figures 4.9, 4.2 and 4.22 that the deviatoric texture, T_d^p, the valence, r^p, and the phase stress ratio, r_σ^p, for each phase remain approximately constant at the axial strain $E_{11} = 5\%$. However, the local porosity, ϕ^p, of each phase is not yet completely constant at this level of the axial strain as the sample continues to dilate slightly until $E_{11} = 7\%$. We approximately consider that all the phases reach the critical state at $E_{11} = 5\%$.

	Sample	$W1'$	$W0'$	$W2'$	$S1'$	$S0'$	$S2'$
E_{11}	7.0 %	5.0 %	5.0 %	5.0 %	5.0 %	5.0 %	5.0 %
P_v^p		0.19	0.13	0.11	0.35	0.15	0.07
T_d^p	0.33	0.17	0.0	-0.09	0.29	0.01	-0.05
ϕ^p	0.21	0.25	0.22	0.21	0.20	0.18	0.17
r^p	4.5	4.2	4.0	3.8	5.3	5.0	4.6
r_σ^p	0.31	0.31	0.28	0.23	0.39	0.27	0.16

Table 4.5. *Values of different characteristics of sample A and of the six phases at the critical state*

Table 4.5 shows some characteristics of the six phases and of sample A at the critical state. We can see that the volume fraction, P_v^p and the deviatoric texture, T_d^p, are close to the respective values at the peak state shown in Table 4.4. Nevertheless, the phases at this state have a higher porosity and a higher valence than at the peak state. This is similar to what is observed for the sample, for which the porosity, Φ, and the coordination number, $\overline{N}$, which is related to mean valence, R, are highest at the critical state (Figures 4.1 and 4.2). The phase stress ratio, r_σ^p, for each phase at this state is significantly lower than the respective value at the peak state. One can remark that the local density and the phase stress ratio, r_σ^p, is still dependent on the phase orientation at the critical state.

	Sample	$W1'$	$W0'$	$W2'$	$S1'$	$S0'$	$S2'$
Loading stage							
P_v^p		0.26	0.14	0.09	0.32	0.13	0.04
T_d^p	0.14	0.23	0.02	-0.08	0.28	0.01	-0.03
r_σ^p	0.28	0.30	0.22	0.16	0.38	0.22	0.08
Unloading stage							
P_v^p		0.27	0.15	0.08	0.33	0.12	0.04
T_d	-0.15	-0.23	-0.01	0.07	-0.28	-0.01	0.03
r_σ^p	-0.32	-0.32	-0.24	-0.18	-0.4	-0.26	-0.1

Table 4.6. *Values of different characteristics of the sample and of the six phases at the critical state for the loading-unloading cycle performed on sample B*

Some characteristics of the six phases at the critical state for the loading/unloading test performed on sample B are shown in Table 4.6. It is interesting to note that, despite the fact that the considered characteristics (P_v^p, T_d^p and r_σ^p) for each phase at the initial state for the loading are very different from those for the unloading, they reach almost the same value at the critical state for the loading and for the unloading. For the deviatoric texture, T_d^p, and the phase stress ratio, r_σ^p, only their sign at the critical state changes from loading to unloading. This is due to the fact that these characteristics are defined in the fixed axes 1 and 2.

4.5. Relevance of the meso-variables and of the meso–macro change of scale

The change from the meso-scale to the macro-scale is based on the definition of entities at the meso-scale, here called phases. The consistency of their mechanical behavior with respect to the loading direction has been demonstrated in the previous sections. The change of scale is also based upon the defined static and kinematic meso-variables and on the volume occupied by each phase in the sample. The macro-stress and macro-strain tensors can be calculated by a volume-weighted average of meso-stress and meso-strain tensors defined for each phase as follows:

$$\Sigma = \sum_{p=1}^{6} P_v^p \sigma^p, \qquad \delta E = \sum_{p=1}^{6} P_v^p \delta \varepsilon^p. \qquad [4.9]$$

From the stress and strain increment tensors defined by equation [4.9], the macroscopic stress ratio, Q/P and the macroscopic volumetric strain increment, δE_v, can be calculated as:

$$Q/P = \frac{\Sigma_1 - \Sigma_2}{\Sigma_1 + \Sigma_2}, \qquad \delta E_v = \delta E_1 + \delta E_2 \qquad [4.10]$$

Figure 4.24 compares the evolution of the macroscopic stress ratio, Q/P and of the incremental macroscopic volumetric strain, E_v, obtained from the boundary conditions and from the volume-weighted average [4.9]. The calculation from the change of scale process in equation [4.9] shows a very good agreement with the calculation resulting from the sample boundaries for the entire loading/unloading cycle. This definitely proves the relevance of the proposed change of scale at three levels. More precisely, it implies that:

– the meso-scale is an appropriate local scale at which to study the local behavior of granular materials;

– the defined meso-variables are relevant static and kinematic variables, even though the definition of the meso-stress raises some intrinsic issues;

– the macro-stress and macro-strain tensors can be obtained by the proposed change of scale from the knowledge of the meso-stress, the meso-strain and the volume fraction of each phase.

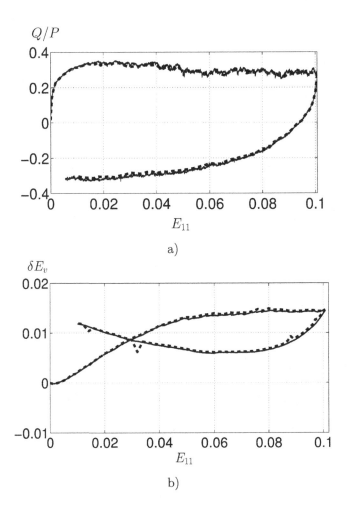

Figure 4.24. *a) Macroscopic stress ratio, Q/P and b) incremental macroscopic volumetric strain, δE_v, defined from the boundary conditions (solid line) and from the volume-weighted average of meso-stress and meso-strain tensors (dashed line) for sample B*

5

Change of Scale Based on Phenomenological Modeling at the Meso-scale

The aim of this chapter is to demonstrate that the knowledge of the mechanical behavior of granular materials at the meso-scale allows the mechanical behavior at the macro-scale to be forecast for any stress path. To this end, we first propose a phenomenological modeling of the mechanical behavior of the phases, which has been analyzed in Chapter 4. Then, we analyze relations between the model parameters defined for each phase and two characteristics of these phases: their orientation with respect to the loading direction $1'$, and their induced anisotropy defined by the component $X_{1'}$ of the anisotropy tensor considered in the phenomenological model. We then propose a simple model to define the evolution of the volumetric amount of each phase related to the loading. Finally, we show that the mechanical behaviors of the considered phases and of the global sample can be forecast on a different stress path from the one used for the identification of the model parameters at the meso-scale. All the analyses presented in this chapter were based on DEM simulations on sample B.

5.1. Phenomenological modeling of the mechanical behavior of phases

In Chapter 4 the mechanical behavior of the six phases, $W1'$, $W0'$, $W2'$, $S1'$, $S0'$ and $S2'$ has been presented. In this section we propose a modeling of this mechanical behavior using a constitutive phenomenological elastic–plastic model called the *CJS model*, which was first developed by Cambou, Jafari and Sidoroff [CAM 87], and used by [CAM 88] to model incomplex loading paths. This model has been improved later for more complex loading paths [MAL 00, BAG 11, DUR 15]. Here, we use a simplified version of this model for 2D media, which considers only two mechanisms: an elastic mechanism and a deviatoric plastic mechanism. Moreover, for the sake of simplicity, the softening behavior after the peak state is not taken into account. This model is presented in Appendix 1.

For each phase p, the incremental strain tensor is written as:

$$\dot{\varepsilon}^{p,\text{tot}} = \dot{\varepsilon}^{p,\text{el}} + \dot{\varepsilon}^{p,\text{pl}}, \qquad [5.1]$$

where $\dot{\varepsilon}^{p,\text{tot}}$ is the total incremental strain tensor, $\dot{\varepsilon}^{p,\text{el}}$ is the incremental elastic strain tensor and $\dot{\varepsilon}^{p,\text{pl}}$ is the incremental plastic strain tensor. It can be pointed out that a kinematic hardening model is considered for the deviatoric mechanism. Therefore, this model allows the initial and induced anisotropies to be taken into account for the calculation of the plastic strain. This is a key point of the considered model.

A precise description of the procedure for the identification of the model parameters is given in Appendix 2. We present in Table 5.1 the initial values of the internal variables and in Table 5.2 the values of the model parameters obtained from the identification procedure performed on the basis of the loading stage. In these tables:

– R^p_{char} is the characteristic radius;

– R^p_{fail} is the failure radius;

– γ^p is the dilatancy parameter;

– b^p and c^p are two parameters involved in the definition of parameter $a^p(E_1)$ that drives the velocity of the kinematic hardening (equation [A1.8]);

– $X^p_{1'\,\mathrm{ini}}$ is the initial value of the component $X^p_{1'}$ of the deviatoric tensor $\boldsymbol{X}^p$ which gives a measure of the initial anisotropy.

Phase	σ_2^p (kPa)	σ_1^p (kPa)	$X^p_{1'\,\mathrm{ini}}$
$\mathcal{W}1'$	109	130	0.0440
$\mathcal{W}0'$	118	123	0.0104
$\mathcal{W}2'$	125	110	-0.0320
$\mathcal{S}1'$	104	158	0.1030
$\mathcal{S}0'$	125	130	0.0098
$\mathcal{S}2'$	142	95	-0.0992

Table 5.1. *Initial values of the stress tensor, σ^p, and of the component $X^p_{1'}$ of the anisotropy tensor, $\boldsymbol{X}^p$, for each phase for the compression loading*

Phase	R^p_{fail}	R^p_{char}	β^p	$b^p\ (10^{-4})$	c^p	θ^p
$\mathcal{W}1'$	0.238	0.100	-0.080	3.2	-14.7	0
$\mathcal{W}0'$	0.198	0.255	-0.300	2.6	-13.1	$\pi/4$
$\mathcal{W}2'$	0.160	0.400	-0.450	2.0	-12.1	$\pi/2$
$\mathcal{S}1'$	0.298	0.010	-0.016	4.0	-19.2	0
$\mathcal{S}0'$	0.198	0.170	-0.400	2.6	-13.1	$\pi/4$
$\mathcal{S}2'$	0.098	0.300	-0.600	1.2	-12.9	$\pi/2$

Table 5.2. *Parameters of the CJS model used for the modeling of the mechanical behavior of each phase for the compression loading. θ^p represents the average orientation of the phase p*

Figure 5.1 shows the evolution of the phase stress ratio, r_σ^p, based on the CJS model and on the parameters presented in Table 5.2. The CJS model provides a correct description of the evolution of the phase stress ratio up to the peak of stresses. After the peak, the phase stress ratio according to the CJS model is always higher than the values given by the DEM simulations, since no softening was considered in this work.

We also show in Figure 5.2 the evolution of the incremental phase strain ratio, r_ε^p. The CJS model seems to give correct results on average despite the difficulty when calibrating model parameters due to the strong fluctuation of the strain. Moreover, no softening for the phase stress ratio towards the critical state was considered in the CJS model, which would have been required in an attempt to retrieve the evolution of r_ε^p towards the critical state in a more precise way. It can be noted that, at the critical state, the incremental volumetric strain is equal to zero for the whole sample, but this is not the case for the considered phases which can show constant, positive, or negative values at the critical state.

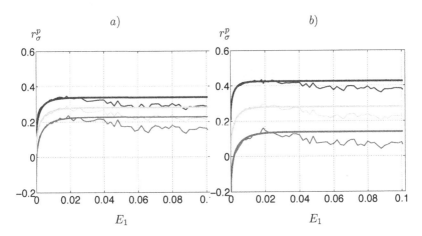

Figure 5.1. *Evolution of the stress ratio: comparison between the model at the meso-scale and the DEM simulations for a) phases $\mathcal{W}$ and b) phases $\mathcal{S}$. Blue lines: meso-domains of orientation class $1'$; Red lines: meso-domains of orientation class $2'$; Green lines: meso-domains of orientation class $0'$; thin lines: DEM; thick lines: model. For a color version of this figure, see www.iste.co.uk/cambou/mesoscale.zip*

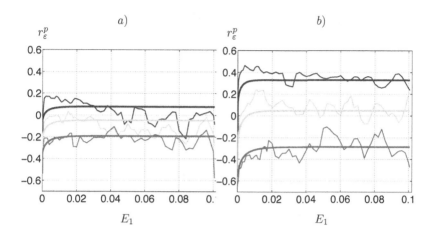

Figure 5.2. *Evolution of the incremental strain ratio: comparison between the model at the meso-scale and the DEM simulations for a) phases $\mathcal{W}$ and b) phases $\mathcal{S}$. Blue lines: meso-domains of orientation class $1'$; Red lines: meso-domains of orientation class $2'$; Green lines: meso-domains of orientation class $0'$; thin lines: DEM; thick lines: model. For a color version of this figure, see www.iste.co.uk/cambou/mesoscale.zip*

5.2. Analysis of the generality of the parameters defined for phases

In Chapter 4, we pointed out that the behavior of each phase was strongly dependent on its orientation with respect to the loading direction. This feature incites us to analyze the consistency between the values of the model parameters for the six phases in more detail. We proposed the hypothesis that the parameters obtained from the identification procedure for each phase (Table 5.2) are only dependent on their internal state (defined by the tensor X^p considered in the CJS model) and on their orientation with respect to the loading direction (defined by the angle θ^p). It can be noted that the orientation of a meso-domain with respect to direction $1'$ corresponds to an angle between two straight lines, so it is within the range $[0; \pi]$. Average values of the phase orientations, θ^p, are equal to 0 for phases $\mathcal{W}1'$ and $\mathcal{S}1'$, $\pi/4$ for phases $\mathcal{W}0'$ and $\mathcal{S}0'$, and $\pi/2$ for phases $\mathcal{W}2'$ and $\mathcal{S}2'$.

Among the eight parameters (see Appendix 2) considered in the CJS model used in this study, some of them do not depend on the considered stress path or on the internal state. This is the case for the two elastic parameters (E and ν) and for the elastic radius of the yielding surface, R_e. The other five parameters analyzed in the following depend on the considered stress path or on the internal state.

5.2.1. Parameter defining the failure surface

Table 5.2 shows that the parameter R_{fail}^p (defining the failure surface) clearly depends on the phase orientation, θ^p. For each elongation class ($\mathcal{W}$ or $\mathcal{S}$), it is possible to define an affine relation between parameters R_{fail}^p and θ^p by relations [5.2] and [5.3]:

– phases $\mathcal{W}$:

$$R_{\text{fail}}^p = -\frac{0.156}{\pi}\theta^p + 0.238, \qquad [5.2]$$

– phases $\mathcal{S}$:

$$R_{\text{fail}}^p = -\frac{0.4}{\pi}\theta^p + 0.298. \qquad [5.3]$$

5.2.2. Parameters defining the plastic hardening law

Parameter b^p depends on the initial anisotropy of phase p, $X_{1'\,\text{ini}}^p$, when a new loading direction is prescribed on the sample. Figure 5.3 allows us to find a quasi affine relationship between parameter b^p and $X_{1'\,\text{ini}}^p$ irrespective of the considered phase:

$$b^p = (11.87 X_{1'\,\text{ini}}^p + 2.58) \times 10^{-4}. \qquad [5.4]$$

Here, we reconsider that $X_{1'\,\text{ini}}^p$ exhibits a positive value if the anisotropy is effectively oriented in the compression direction, $1'$, and a negative value if the direction of the anisotropy is perpendicular to the compression direction.

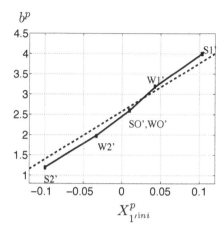

Figure 5.3. *Evolution of the model parameter b^p with the initial anisotropy for the loading. Solid lines: DEM; dashed line: model*

5.2.3. *Parameters characterizing the plastic volume change*

First of all, we point out from Figures 4.17 and 4.18 that at the critical state the volumetric deformation rate is equal to zero for the whole sample, but it is clearly different from zero for each phase.

Phases	θ^p	$r^p_{\varepsilon\,\mathrm{crit}}$
$\mathcal{W}1'$	0	0.098
$\mathcal{W}0'$	$\pi/4$	-0.047
$\mathcal{W}2'$	$\pi/2$	-0.190
$\mathcal{S}1'$	0	0.329
$\mathcal{S}0'$	$\pi/4$	0.048
$\mathcal{S}2'$	$\pi/2$	-0.285

Table 5.3. *The phase deformation ratio at the critical state, $r^p_{\varepsilon\,\mathrm{crit}}$, with respect to the orientation θ^p of each phase*

In Table 5.3 we present the values of the phase strain ratio at the critical state, $r^p_{\varepsilon\,\mathrm{crit}}$, with respect to the orientation of each phase, θ^p. It is clear in this table that $r^p_{\varepsilon\,\mathrm{crit}}$ depends on the orientation θ^p in an affine way. Affine relationships can thus be drawn for the weak and the strong phases:

– phases $\mathcal{W}$:

$$r^p_{\varepsilon\,\text{crit}} = -\frac{0.578}{\pi}\theta^p + 0.098, \qquad\qquad [5.5]$$

– phases $\mathcal{S}$:

$$r^p_{\varepsilon\,\text{crit}} = -\frac{1.177}{\pi}\theta^p + 0.329. \qquad\qquad [5.6]$$

This analysis provides evidence that the evolution of ε^p_v is different from one phase to another and that the parameters β^p and R^p_{char} then depend on the orientation θ^p.

Parameter R^p_{char} corresponds to the radius of the characteristic surface (Figure A1.1). This parameter is defined from the knowledge of the characteristic state which is the stress state corresponding to $\dot{\varepsilon}^p_v = 0$ in the initial part of the stress–strain curve. Parameter β^p is usually defined from the knowledge of the maximal slope of the evolution of ε^p_v with respect to E_1. By analyzing the results of the DEM simulations in Chapter 4, it can be noted that phases $\mathcal{S}1'$ and $\mathcal{W}1'$ are always dilative and phases $\mathcal{S}2'$, $\mathcal{W}2'$ and $\mathcal{W}0'$ are always contractive. Then, a clear and straightforward process to identify R^p_{char} is impossible for these phases. Appendix 2 presents the procedure which has been used to estimate parameters β^p and R^p_{char} presented in Table 5.2.

5.3. Forecast of the mechanical behavior of the phases on a new loading path

The validation of the considered model needs to perform a forecast of the mechanical behavior on a loading path which is different from the path used for the identification process. We have chosen to perform this forecast on an unloading, which has been numerically performed from the critical state achieved at the end of the compression path.

This unloading path is very different from the initial compression test realized on an isotropic sample, since the sample is highly anisotropic at the beginning of the unloading. For this unloading, the compression direction as defined in the Introduction is direction $1'$ which corresponds to the fixed direction 2. Orientations of the phases have thus to be defined with respect to this new meaning of the direction $1'$.

The associated model parameters of each phase can be computed using relations [5.2], [5.3] and [5.4] and Table 5.2 with the new values of θ^p and $X^p_{1'\,\text{ini}}$. It can be noted that at the beginning of the unloading, the major principal direction of the anisotropy is no longer oriented in the new compression direction, and thus $X^p_{1'\,\text{ini}}$ exhibits a negative value.

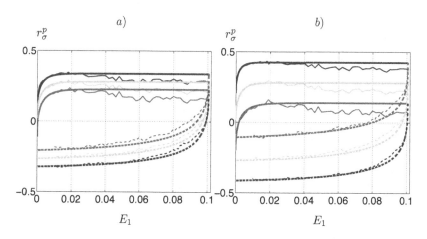

Figure 5.4. *Validation of the proposed model for the phase stress ratio: a) phases $\mathcal{W}$ and b) phases $\mathcal{S}$. Blue lines: meso-domains of orientation class $1'$; red lines: meso-domains of orientation class $2'$; green lines: meso-domains of orientation class $0'$; thin lines: DEM; thick lines: model. Solid lines: loading; dashed lines: unloading. For a color version of this figure, see www.iste.co.uk/cambou/mesoscale.zip*

Figure 5.4 presents the prediction of the phase stress ratio for phases $\mathcal{W}$ and $\mathcal{S}$. It can be seen that the proposed model gives a correct prediction of the stress ratio, r^p_σ, at the beginning and at the end of the unloading stage. In the former case, it mainly results from a correct prediction of parameter b^p related to the stiffness of the material in that direction at the earlier stage of the loading. In the latter case, it is due to a correct estimate of the failure radius, R_{fail}^p.

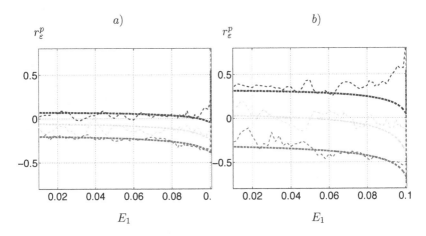

Figure 5.5. *Validation of the proposed model for the phase strain ratio: a) phases $\mathcal{W}$ and b) phases $\mathcal{S}$. Blue lines: meso-domains of orientation class $1'$; red lines: meso-domains of orientation class $2'$; green lines: meso-domains of orientation class $0'$; thin lines: DEM; thick lines: model. Solid lines: loading; dashed lines: unloading. The figures must be read from right to left. For a color version of this figure, see www.iste.co.uk/cambou/mesoscale.zip*

Figure 5.5 presents the prediction of the phase strain ratio for the phases $\mathcal{W}$ and $\mathcal{S}$. It can be noted that for the final part of the unloading stage (deformation E_1 ranging from 1% to 5%), the prediction is fairly correct compared to the DEM result, which is not the case for the early stage of the unloading. More precisely, if in direction $2'$ the prediction seems fairly correct, it tends to get worse as the orientation of the phase gets closer to direction $1'$. This may be due to a poor modeling of the elastic behavior which is predominant at the beginning of the unloading. In fact, in the proposed model the elasticity of each phase is assumed to be isotropic, while it is well known that the elasticity of granular materials is anisotropic [BAR 94, TAM 05]. This can explain why it can be seen from the DEM simulations that the stress-strain curve for each phase at the beginning of the unloading depends on the direction of the anisotropy with respect to the compression direction, $1'$, with a greater dependency as the anisotropy level increases (phases S are more anisotropic than phases W).

5.4. Change of scale: from the mesoscopic phases to the macroscopic sample

In the three previous sections, the modeling of behavior of the granular phases has been proposed from an identification on a loading path and validated from a prediction on an unloading path. In this section, we present the procedure that allows the behavior of granular materials at the RVE-scale to be defined from the constitutive model defined for each phase at the meso-scale.

5.4.1. Modeling of the evolution of the volumetric percentage of each phase

Figure 5.6 shows the evolution of the volumetric percentage of each phase throughout loading (solid lines). The volume of the phases oriented in the compression direction, $1'$, increases while the volume of the phases oriented in the extension direction, $2'$, decreases. It is also clear that the volumetric amount of each phase, P_v^p, reaches a plateau corresponding to the critical state. Therefore, the following non-linear incremental model has been proposed for the modeling of the incremental change in the volumetric percentage of each phase:

$$\dot{P}_v^p = \gamma^p (P_{v\,\mathrm{crit}}^p - P_v^p(E_{1'})). \qquad [5.7]$$

The two parameters characterizing this model, γ^p and $P_{v\,\mathrm{crit}}^p$, can easily be defined from results obtained when simulating a compression test in direction $1'$ (Figure 5.6). $P_{v\,\mathrm{crit}}^p$ corresponds to the limit value observed at high deformation and γ^p can be calculated from the initial slope of the curves in Figure 5.6.:

$$\gamma^p = \frac{\dot{P}_v^p(0)}{P_{v\,\mathrm{crit}}^p - P_v^p(0)}. \qquad [5.8]$$

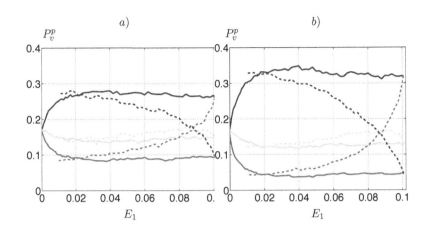

Figure 5.6. *Percentage by volume occupied by the six phases during the loading–unloading test: a) phase of elongation class W and b) phases of elongation class S. Blue lines: meso-domains of orientation class* 1'; *red lines: meso-domains of orientation class* 2'; *green lines: meso-domains of orientation class* 0'; *black lines: sample. Solid lines: loading path; dashed lines: unloading path. For a color version of this figure, see www.iste.co.uk/cambou/mesoscale.zip*

The initial values of P_v^p, $P_v^p(0)$, the initial slopes of the curves, $\dot{P}_v^p(0)$, and the final values of P_v^p, $P_{v\,\mathrm{crit}}^p$, obtained from Figure 5.6. are presented in Table 5.4. Figure 5.7 also shows that the model proposed by equation [5.7] gives results in good agreement with the DEM numerical simulation.

We will now analyze the ability of the proposed model to describe the evolution in terms of volumetric amount of each phase with the parameters obtained above. We remember that the identification of these parameters is based on the first loading in the compression direction, 1'. For this purpose, we consider the relationship that could exist between parameter γ^p and the internal state variable X^p considered in the elastic–plastic model. Figure 5.8 shows that γ^p is an affine function of $|X_{1'\,\mathrm{crit}}^p - X_{1'\,\mathrm{ini}}^p|$:

$$\gamma^p = (-1.25|X_{1'\,\mathrm{crit}}^p - X_{1'\,\mathrm{ini}}^p| + 0.323) \times 10^{-3}. \qquad [5.9]$$

Phase	$X^p_{1'\,\text{ini}}$	$X^p_{1'\,\text{crit}}$	$|X^p_{1'\,\text{crit}} - X^p_{1'\,\text{ini}}|$	$P^p_v(0)$	$P^p_{v\,\text{crit}}$	$\gamma^p\ (\times 10^4)$
$\mathcal{W}1'$	0.0440	0.1384	0.0944	0.17	0.27	2
$\mathcal{W}0'$	0.0104	0.1100	0.0996	0.17	0.14	2
$\mathcal{W}2'$	-0.0319	0.0831	0.115	0.17	0.09	1.9
$\mathcal{S}1'$	0.1030	0.1807	0.0777	0.17	0.33	2.27
$\mathcal{S}0'$	0.0098	0.1100	0.1002	0.16	0.12	2
$\mathcal{S}2'$	-0.0992	0.0392	0.1384	0.15	0.04	1.5

Table 5.4. *Values of the parameters characterizing the model defining the volume evolution of the phases and corresponding values of the internal state variable X^p. These values are based on the initial compression test*

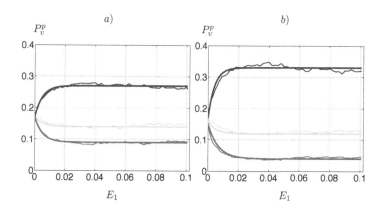

Figure 5.7. *Comparison between the measured values of P^p_v in the DEM simulations and the modeling of P^p_v given by equation [5.7] for a) phases $\mathcal{W}$ and b) phases $\mathcal{S}$. Blue lines: meso-domains of orientation class $1'$; red lines: meso-domains of orientation class $2'$; green lines: meso-domains of orientation class $0'$; thin lines: DEM; thick lines: model. For a color version of this figure, see www.iste.co.uk/cambou/mesoscale.zip*

Equation [5.9] can be used for any loading direction. For the validation procedure, Table 5.5 shows the values of the parameters and of the internal state variables used to model the evolution in volumetric percentage of each phase throughout the unloading path. Figure 5.9 shows the comparison between the result obtained from the DEM simulations and the result obtained from the proposed model.

A very good agreement is observed between the numerical simulations and the proposed model for the evolution of P_v^p during the unloading.

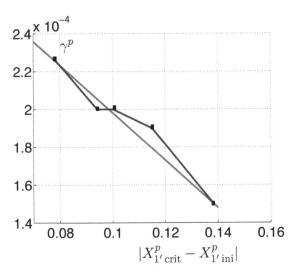

Figure 5.8. *Relation between parameter γ^p and the internal variable X^p used in the elastic-plastic model. Blue line: DEM; red line: model. For a color version of this figure, see www.iste.co.uk/ cambou/mesoscale.zip*

Phase	$X_{1'\,\mathrm{ini}}^p$	$X_{1'\,\mathrm{crit}}^p$	$\left\lvert X_{1'\,\mathrm{crit}}^p - X_{1'\,\mathrm{ini}}^p \right\rvert$	$P_v^p(0)$	$P_{v\,\mathrm{crit}}^p$	$\gamma^p\,(\times 10^4)$
$\mathcal{W}1'$	0.0831	-0.1384	0.22	0.09	0.27	0.47
$\mathcal{W}0'$	0.0110	-0.1100	0.22	0.14	0.14	0.47
$\mathcal{W}2'$	0.1384	-0.0831	0.22	0.27	0.09	0.47
$\mathcal{S}1'$	0.0392	-0.1807	0.22	0.04	0.33	0.47
$\mathcal{S}0'$	0.1100	-0.1100	0.22	0.12	0.12	0.47
$\mathcal{S}2'$	0.1807	-0.0392	0.22	0.33	0.04	0.47

Table 5.5. *Predicted values of the parameters characterizing the model defining the volume evolution of the phases, and corresponding values of the internal state variable X^p considered for the forecast performed for the unloading stress path*

5.4.2. Achievement of the meso–macro change of scale

Finally, the stress tensor $\Sigma(E_{1'})$ and the strain tensor $E(E_{1'})$ at the macro-scale are derived from the information for each phase at the meso-scale using the following relationships:

$$\Sigma(E_{1'}) = \sum_{p=1}^{6} \sigma^p(E_{1'}) P_v^p(E_{1'}), \qquad [5.10]$$

$$\varepsilon(E_{1'}) = \sum_{p=1}^{6} \varepsilon^p(E_{1'}) P_v^p(E_{1'}), \qquad [5.11]$$

where p is one of the six phases considered and $P_v^p(E_{1'})$ is the volumetric percentage occupied by this phase p at the deformation state $E_{1'}$. Relations [5.10] and [5.11] are valid for any stress path as long as the directions of the principal stresses do not rotate, which is the case here. Therefore, to emphasize the validity of our approach, the macro-stress for the unloading presented in Chapter 2 is predicted from the knowledge of the behavior of the six phases throughout the loading and of the volumetric amount occupied by each phase.

Figure 5.10 shows that the prediction of the unloading path is in very good agreement with the numerical DEM simulations. This result points out the potential and the efficiency of the proposed change of scale procedure. This procedure is valid for any cyclic biaxial loading. Moreover, it is worth mentioning that *the prediction of any cycle just requires the identification of the model parameters on a first (monotonous) loading stage.*

It can be emphasized that the considered CJS model is able to give correct forecasts even for complex loadings such as loading paths with rotation of directions of the principal stresses. A generalization of the proposed modeling is thus possible, but more specific analyses are required to define specific correlations between parameters considered for the modeling of the behavior of phases and the direction of loading path in a very general case.

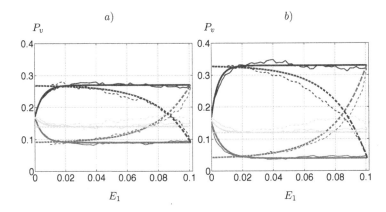

Figure 5.9. *Comparison of the evolution of the volume of phases from the DEM simulations and from the proposed model (equation [5.7]) for a) phases $\mathcal{W}$ and b) phases $\mathcal{S}$. Blue lines: meso-domains of orientation class $1'$; red lines: meso-domains of orientation class $2'$; green lines: meso-domains of orientation class $0'$; thin lines: DEM; thick lines: model. Full-lines: loading; dashed lines: unloading. For a color version of this figure, see www.iste.co.uk/cambou/mesoscale.zip*

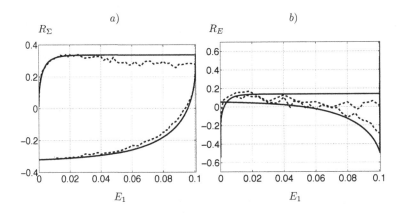

Figure 5.10. *Comparison of the results of the DEM simulations and of the modeling based on the change of scale for the loading (identification) and unloading (forecast) paths at the REV scale: a) macroscopic stress ratio and b) macroscopic strain ratio. Dashed lines: DEM; solid lines: model*

Conclusions and Prospects

This book was aimed firstly at presenting an accurate analysis of 2D granular material at an intermediary scale, here called *the meso-scale*. This analysis was focused essentially on local variables of texture, on stress, and on strain tensors defined at this intermediary scale. The second purpose of this book was to present a multi-scale approach considering this meso-scale to model the mechanical behavior of granular materials. At this scale, our analyses were based on numerical DEM simulations of biaxial tests performed on 2D granular materials. The studied granular materials were split into local structures, called *meso-domains*, defined by closed loops of contacting particles surrounding a void volume (rattlers are disregarded). We first focused our work on the analysis of different variables defined at the meso-scale: texture variables defining the internal state, meso-strain and meso-stress (item 1). We then introduced the concept of *phases*, defined at the meso-scale as clusters of meso-domains sharing the same texture characteristics (item 2, 3, and 4). Using an identification procedure, we fitted an elastic plastic model (a simplified version of the CJS model) to the behavior of each phase given by the DEM–simulations (items 5). Finally, we analyzed the ability to perform a change of scale in granular materials knowing both the volumetric amount and the behavior of each phase (item 5).

1) We first clarified the different meso–variables allowing internal state, strain, and stress to be defined for each meso-domain. The internal state of a meso-domain can be defined by a meso-porosity, a meso-domain valence, and a loop tensor. This tensor allows an elongation ratio and an orientation to be defined for each meso-domain. The knowledge of the meso-domain orientation allows the anisotropy of a cluster of meso-domains to be easily defined. The average strain in each meso-domain is defined from the displacements of the vertices of the considered domain. The meso–stress is defined from the forces acting on the particles located on the boundary of a given meso-domain. This definition is based on an assumption and leads to a small discrepancy between the average stress computed from the meso–stress and the value defined at the sample boundaries. This discrepancy nevertheless remains negligible if the meaningful stress ratio, Q/P, is considered.

2) By analyzing different numerical simulations of biaxial loading and unloading paths, we pointed out that the distribution of local strains and stresses were highly correlated with the internal state of meso-domains characterized by their elongation ratios and their orientations. To get a more accurate knowledge of this correlation we considered six *phases* which are clusters of meso-domains sharing similar texture characteristics: the elongation degree and the orientation. The analysis of the internal states of these phases led to the following conclusions:

i) During both loading and unloading paths, meso-domains tend to be preferentially oriented in the major principal compression direction, $1'$. The distribution of the strongly elongated meso-domains (elongation class S) is clearly much more anisotropic than that of the weakly elongated meso-domains (elongation class W).

ii) The analysis of the texture tensor of the six considered phases provided evidence that whatever the loading direction, the phases oriented in the major principal compression direction, $1'$, develop a high level of anisotropy. Moreover, this level is much higher than the one developed by the phases oriented in the extension direction, $2'$.

iii) The six considered phases have different values of porosity at the initial isotropic state: meso-domains of the strong elongation class, S, are looser than the sample and meso-domains of the weak elongation class, W, are denser than the sample. Whatever the loading direction, the porosity of the meso-domains oriented in the major principal compression direction (phases $W1'$ and $S1'$) can bear a much higher dilatancy than the meso–domains oriented in the extension direction (phases $W2'$ and $S2'$).

iv) Whatever the loading direction, the volume occupied by the phases oriented in the major principal compression direction ($W1'$ and $S1'$) increases greatly and rapidly, reaches a maximum and then decreases slightly, while the volume occupied by the phases oriented in the extension direction ($W2'$ and $S2'$) decreases quickly and tends to a constant value.

3) During loading and unloading paths, the distributions of the local strain and stress tensors are highly dependent on the local characteristics.

i) Stress state in each phase is characterized by the phase stress ratio. At the initial isotropic stress state, the phases oriented in direction $1'$ exhibit positive values of the phase stress ratio, while the phases oriented in direction $2'$ exhibit negative values of this ratio. During loading and unloading, the phase stress ratio increases more in the phases oriented in direction $1'$ than in the other phases. The phase stress ratio of each phase stabilizes when reaching the macroscopic critical state. With the same orientation, the strongly elongated phases (elongation class S) develop higher values of the phase stress ratio than the weakly elongated phases (elongation class W).

ii) The DEM simulations clearly show that the strain of each phase depends considerably on the phase orientation: the phases oriented in the major principal direction, $1'$, dilate continuously, while the phases oriented in the extension direction, $2'$, contract continuously, particularly for the strongly elongated phases. These conclusions are consistent with the evolution of both volume fraction and stress state of each phase. The phases oriented in the compression direction $1'$ are able to dilate strongly, which allows them to resist a high deviatoric stress state; and hence their volume increases to support the increasing deviatoric stress. On the contrary, the phases oriented in the extension direction $2'$ are not, due to their weak dilatancy, able to support a high deviatoric stress state; and hence their volume decreases. Finally, we pointed out that the phases oriented in an intermediate direction ($W0'$ and $S0'$) are neutral in terms of influence.

4) Using local information at the meso-scale we computed the strain and stress tensors in all meso-domains, and also their volumetric averages along the considered loading and unloading. We plotted stress–strain curves at the sample level obtained from the averaging procedure and then compared these curves with the stress-strain curves obtained from the boundary conditions. The strong agreement between these curves demonstrated the relevance of the defined meso-variables: the meso-strain and meso-strain tensors.

5) In Chapter 4, we showed that the mechanical behavior of the *phases* depend clearly on their internal state. This demonstrates the relevance of the concept of *phases* defined from local meso-domain characteristics. In Chapter 5 we then analyzed the possibility of considering that these phases defined at a meso-scale propose a new method of performining a change of scale for the definition of the mechanical behavior of granular materials.

i) The mechanical behavior of the six considered phases was modeled by an elastic-plastic model based on a deviatoric plastic mechanism with a kinematic hardening. Moreover, the evolution of the volumetric percentage of the phases was defined by a simple model. The parameters used in these models for each phase were identified based on the loading stage of a biaxial test.

ii) These parameters were found to be strongly dependent on the initial anisotropy and on the orientation of the considered phase with respect to the compression direction, $1'$. These general relationships provide the set of parameters with a more general validity irrespective of the stress path.

iii) On the basis of the identification of the model parameters based on the compression loading, a prediction of the mechanical behavior of the phases was performed on a different stress path: an unloading biaxial path. These predictions make use of the parameters identified for each phase during the loading path. The obtained results fitted well with the results given by the DEM numerical simulations, except for the strain tensor at the beginning of the unloading path. This latter feature may be due to the nature of elasticity in the chosen constitutive model which is isotropic for the sake of simplicity. In granular media, this elasticity is known to be anisotropic.

iv) The overall behavior of a Representative Elementary Volume on an unloading biaxial path was defined from the modeling of the mechanical behavior of the six considered phases and of their changes in volume. This overall behavior was compared to the results obtained from the DEM simulation. The agreement between the predicted unloading and the DEM results proves the relevance of the defined meso-variables and also the relevance of the change of scale approach considered in this book based on analyses at the meso-scale.

Finally, this study not only allows a very clear description of the physical evolution of a granular assembly at the meso-scale to be proposed, but also demonstrates the great interest of using such an intermediary scale and the concept of *phases* in a change of scale approach. Future research is needed to develop a 3D constitutive model based on this new concept. This approach is interesting because the mechanical behavior of each phase is defined by only two parameters: the orientation with respect to the loading direction and the initial anisotropy. There is no need for other internal variables to define the phase behaviors for any kind of loading paths. However, the resulting mechanical behavior at the VER-scale is dependent on the internal state and on the loading path since the volumetric amount of each phase is dependent on the two latter ones. Parameters of this 3D model defined for each phase will be obtained from the global behavior and from relationships between these parameters and the material state of each phase. These relationships would be obtained using DEM simulations. The basic hypothesis in this approach is that these relationships obtained from DEM simulations could be used for actual materials.

APPENDICES

Appendix 1

The Simplified Elastic Plastic CJS Model for 2D Materials

In this work, a simple form of the elastic plastic CJS model [CAM 88a, MAL 00, BAG 11, DUR 15] is used for 2D granular materials. As presented in equation [5.1], the total strain is separated into an elastic part and a plastic part. For the sake of simplicity, a linear elasticity is considered for the modeling of elastic strains. The incremental elastic strain tensor is calculated by Hooke's law, using a Young's modulus, E, and a Poisson's ratio, ν. Plasticity is generated when the stress state reaches the deviatoric yield criterion. Nevertheless, the stress state must be contained within a valid domain of stresses whose boundary is defined by the failure surface, which is isotropic and defined as:

$$f(\boldsymbol{\sigma}) = s_{II} - R_{\text{fail}}I_1, \qquad [\text{A1.1}]$$

where R_{fail} is the radius of the failure surface, s_{II} is the second invariant of the deviatoric part s of the stress tensor $(s_{II} = \sqrt{s_{ij}s_{ij}})$ and I_1 is the first invariant of the stress tensor $\boldsymbol{\sigma}$ $(I_1 = \sigma_{kk})$.

The deviatoric plastic mechanism is defined by the following yield criterion, which exhibits the same shape as the failure surface (Figure A1.1):

$$f(\boldsymbol{\sigma}, \boldsymbol{X}) = q_{II} - R_e I_1, \qquad [A1.2]$$

where q_{II} is the norm of tensor $\boldsymbol{q}$ defined by:

$$\boldsymbol{q} = \boldsymbol{s} - \boldsymbol{X} I_1. \qquad [A1.3]$$

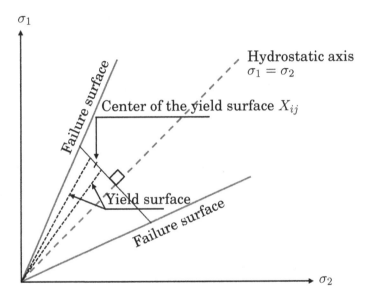

Figure A1.1. *Definition of the failure surface and of the yield surface of the CJS model.*

In relations [A1.2] and [A1.3], $\boldsymbol{X}$ is a deviatoric tensor associated with the center of the deviatoric yield surface: it defines the anisotropy state of the considered material, R_e is the elastic radius which defines the amplitude of the elastic domain and is a model parameter. In this study, this elastic parameter was considered very small since experiments showed that granular materials exhibit a very small actual elastic domain.

The center of the yield surface may change according to the kinematic hardening mechanism:

$$\dot{X} = a\dot{\alpha} \quad \text{with} \quad \dot{\alpha} = \lambda^d I_1 \left(\frac{\dot{q}}{q_{II}} - \phi \dot{X} \right) \left(\frac{I_1}{2p_a} \right)^m , \quad [A1.4]$$

where:

– a is the model parameter ruling the velocity of the kinematic hardening;

– α is the tensor related to the kinematic hardening variable X;

– ϕ is a function which limits the evolution of the kinematic hardening;

– the term $\left(\dfrac{I_1}{2p_a} \right)^m$ reduces the influence of the mean pressure in the mechanical behavior (p_a is a reference pressure set to 100 kPa and $m = 1.2$).

Function ϕ is defined as:

$$\phi = \phi_0 q_{II} \quad \text{with} \quad \phi_0 = \frac{1}{R_{\text{fail}}}. \qquad [A1.5]$$

The increment of the plastic volumetric strain related to the plastic deviatoric mechanism is calculated by:

$$\dot{\varepsilon}_v^{dp} = \beta \left(\frac{s_{II}}{s_{II}^{\text{char}}} - 1 \right) \frac{|s : \dot{e}^{dp}|}{s_{II}}, \qquad [A1.6]$$

where β is the dilatancy parameter, e^{dp} is the deviatoric part of the strain tensor and s_{II}^{char} is the value of s_{II} at the characteristic state for the current mean pressure. The characteristic surface, which is also an isotropic surface, with the same shape as the failure surface, is written as:

$$f^{\text{char}}(\boldsymbol{\sigma}) = s_{II}^{\text{char}} - R_{\text{char}}I_1, \qquad [\text{A1.7}]$$

where R_{char} is the radius of the characteristic state and is a model parameter. Considering relation [A1.7] contractancy occurs if $s_{II} < s_{II}^{\text{char}}$ and dilatancy occurs otherwise.

In the initial CJS model the stress ratio curve at the beginning of loading and unloading is not well described. So we have improved the model by considering that the kinematic hardening parameter a at a given state depends on the induced anisotropy X at this state. The parameter a is then defined using the following exponential relationship:

$$a = be^{c(X_1 - X_1 \, \text{ini})}, \qquad [\text{A1.8}]$$

where X_1 is the current anisotropy, $X_1 \, \text{ini}$ is the initial anisotropy for the considered loading path (e.g loading or unloading) and b and c are two parameters of the kinematic hardening.

Appendix 2

Calibration of the Model Parameters

The identification of the model parameters is processed on a monotonous loading path through the following procedure for each phase:

– *Elastic parameters*: the elastic parameters (Young's modulus and Poisson's ratio) are identical for all the phases. Young's modulus can be determined from the stiffness of the particles by the following relation proposed by [CAM 95] for 2D materials:

$$E = \frac{2k_n}{\pi}\frac{1+\eta}{3+\eta},$$ [A2.1]

where k_n is the normal particle stiffness and η is the ratio of the tangential stiffness k_s to the normal stiffness k_s. In this work, $\eta = 1$, $k_n = 10^9$ N/m, which gives a Young's modulus equal to 3.18×10^8 N/m. The Poisson's ratio is supposed to be equal to 0.3.

– *Elastic radius*: the elastic radius, R_e, defines the amplitude of the elastic domain. In granular materials, this domain is typically very small (corresponding to a deviatoric deformation smaller than 10^{-5} [TAT 92]). This value does not greatly influence the results of the modeling. For the sake of simplicity, a default value of 0.014 was chosen for all phases.

−*Failure radius*: this parameter is determined by considering the maximum value of the stress ratio r_σ^{max} observed in the DEM simulations. It is different for each phase. This parameter can be written as:

$$R_{\text{fail}} = \frac{s_{II}^{\text{fail}}}{I_1^{\text{fail}}} = \frac{s_1^{\text{fail}}\sqrt{2}}{I_1^{\text{fail}}} = \frac{r_\sigma^{max}\sqrt{2}}{2}. \qquad \text{[A2.2]}$$

− *Center of the yield surface at the initial state*: we suppose that the center of the yield surface corresponds to the initial state of stress for each phase in the DEM simulation. We saw in Chapter 4 that the initial state of stress for a given phase is anisotropic although the initial macroscopic stress is isotropic. The deviatoric tensor X is then deduced for each phase from the initial value of the major principal deviatoric stress, s_1^0, and the initial isotropic stress, I_1^0.

$$X_1^0 = \frac{s_1^0}{I_1^0}, \quad X_2^0 = -X_1^0. \qquad \text{[A2.3]}$$

−*Parameters of the kinematic hardening mechanism*: Parameters b and c in equation [A1.8] have to be identified for each phase. First, the value of b is identified in order to provide a correct slope for the evolution of the stress ratio at the beginning of the deviatoric loading (Figure A2.1(a)). For this range of deformation, the current anisotropy is very close to the initial anisotropy. Therefore, the exponential term involved in equation [A1.8] is close to 1 and then $a(E_1) \approx b$.

Then, parameter c is identified by a trial-and-error method in order to provide a correct curvature for the evolution of the stress ratio in the range of deformations prior to the maximum stress ratio (e.g axial strain ranging generally between 0.5% and 2%). The final result is showed in Figure A2.1(b).

−*Characteristic radius R_{char} and dilatancy parameter β*: parameter R_{char} corresponds to the radius of the characteristic surface (Figure A1.1). It is defined from the knowledge of the characteristic state which corresponds, in the initial part of the stress–strain curve, to $\dot{\varepsilon}_v = 0$. Parameter β is usually defined from knowledge of the maximal slope of the evolution of the volumetric strain ε_v with respect to E_1. By analyzing the results of the DEM simulation (Chapter 4), it can be observed that the phases $S1'$ and $W1'$ are always dilative and the phases $S2'$, $W2'$ and $W0'$ are always contractive. Thus, a clear and straightforward process to identify these parameters is impossible for these phases. So it is possible to define accurate values of β and R_{char} only for the phases which pass through the characteristic state. However, as mentioned in Chapter 4 some phases do not pass through the characteristic state. For these phases, the characteristic radius R_{char} is defined by the following rule: if the phase is always dilative, then $R_{\mathrm{char}} < R_{\mathrm{fail}}$; if the phase is always contractive, then $R_{\mathrm{char}} > R_{\mathrm{fail}}$. The incremental volumetric strain is defined by equation [A1.6] involving β and R_{char} (through equation [A1.7]). Considering this equation for biaxial tests leads to the following equation:

$$r_\varepsilon = \beta \left(\frac{R_{\mathrm{fail}}}{R_{\mathrm{char}}} - 1 \right) \sqrt{2}, \qquad\qquad \text{[A2.4]}$$

where $r_\varepsilon = \dot{\varepsilon}_v / \dot{\varepsilon}_d$ is the strain ratio.

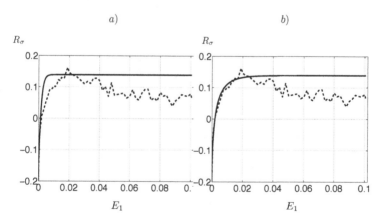

Figure A2.1. *Identification of the parameters of the hardening law for the phase $S2'$. a): parameter b; b): parameter c). Solid lines: model; Dashed-lines: DEM*

Thus, for the phases which do not pass through the characteristic state which corresponds to $r_\varepsilon = 0$, it is possible to use equation [A2.4] and the rules defined above. Equation [A2.4] is considered in a zone of axial deformations ranging between 2% and 4%. It allows us to retrieve the average value for the evolution of the strain ratio observed in the DEM simulation.

Bibliography

[AL 03] AL-RAOUSH R., THOMPSON K., WILLSON C.S., "Comparison of network generation techniques for unconsolidated porous media", *Soil Science Society of America Journal*, vol. 67, no. 6, pp. 1687–1700, 2003.

[AUV 92] AUVINET G., "Stress transfer within granular geomaterials", in LUTES L.D. (ed.), *Engineering Mechanics*, ASCE, New York, pp. 159–162, 1992.

[BAC 92] BACCONNET C., GOURVES R., "Random aspect of the stress inside granular media", in LUTES L.D. (ed.), *Engineering Mechanics*, ASCE, New York, pp. 163–166, 1992.

[BAG 93] BAGI K., "On the definition of stress and strain in granular assemblies through the relation between micro- and macro-level characteristics", in THORNTON C. (eds), *Proceedings of Powder and Grains 1993*, Balkema, pp. 117–121, 1993.

[BAG 96] BAGI K., "Stress and strain in granular assemblies", *Mechanics of Materials*, vol. 22, no. 3, pp. 165–177, 1996.

[BAG 06] BAGI K., "Analysis of microstructural strain tensors for granular assemblies", *International Journal of Solids and Structures*, vol. 43, no. 10, pp. 3166–3184, 2006.

[BAG 11] BAGAGLI Y., Modélisation cyclique des sols et interfaces sol/structure, PhD thesis, Ecole Centrale de Lyon, 2011.

[BAR 94] BARDET J.P., "Numerical simulations of the incremental responses of idealized granular materials", *International Journal of Plasticity*, vol. 10, no. 8, pp. 879–908, 1994.

[BAZ 90] BAZANT Z.P., OZBOLT J., "Non-local microplane model for fracture, damage and size effects in structures", *Journal of Engineering Mechanics*, vol. 116, no. 11, pp. 2485–2505, 1990.

[CAI 95] CAILLERIE D., "Evolution quasistatique d'un milieu granulaire, loi incrémentale par homogénéisation", in *Des géomatériaux aux ouvrages*, Hermès, 1995.

[CAI 02] CAILLERIE D., CAMBOU B., "Les techniques de changement d'échelle dans les matériaux granulaires", in CAMBOU B., JEAN M. (eds), *Micromécanique des milieux granulaires*, pp. 73–198, Hermes Science, 2002.

[CAM 88a] CAMBOU B., JAFARI K., "A constitutive model for granular materials based on two plasticity mechanisms", in SAADA A., BIANCHINI G. (eds) *Proc. Int. Symp. Constitutive Equations for Granular and Non-cohesive Soils*, Balkema, pp. 149–167, 1988.

[CAM 88b] CAMBOU B., LANIER J., "Induced anisotropy in cohesionless soil: Experiments and modelling", *Computers and Geotechnics*, vol. 6, no. 4, pp. 291–311, 1988.

[CAM 95] CAMBOU B., DUBUJET P., EMERIAULT H. *et al.*, "Homogenization for granular materials", *European Journal of Mechanics A/Solids*, vol. 14, no. 2, pp. 255–276, 1995.

[CAM 00] CAMBOU B., CHAZE M., DEDECKER F., "Change of scale in granular materials", *European Journal of Mechanics, A/Solids*, vol. 19, no. 6, pp. 999–1014, 2000.

[CAM 10] CAMBOU B., JEAN M., RADJAÏ F., *Micromechanics of Granular Materials*, ISTE Ltd, London and John Wiley & Sons, New York, 2010.

[CAM 12] CAMBOU B., JEAN M., RADJAÏ F., *Matériaux granulaires : modélisation et simulation numérique*, Hermes Science-Lavoisier, 2012.

[CAM 14] CAMBOU B., MAGOARIEC H., VINCENS E., "State internal variables at different scales for the modeling of the behavior of granular materials", *Continuum Mechanics and Thermodynamics*, vol. 27, nos. 1–2, pp. 223–241, 2014.

[CAS 40] CASAGRANDE A., "Characteristics of cohesionless soils affecting the stability of slopes and earth fills", *Journal of Boston Society of Civil Engineers*, pp. 257–276, 1940.

[CHA 76] CHAPUIS R., De la structure géométrique des milieux granulaires en relation avec leur comportement mécanique, PhD thesis, Ecole Polytechnique de Montréal, 1976.

[CHA 90] CHANG C.S., MISRA A., "Application of uniform strain theory to heterogeneous granular solids", *Journal of Engineering Mechanics*, vol. 116, no. 10, pp. 2310–2328, 1990.

[CHA 94] CHANG C.S., LIAO C., "Estimates of elastic modulus for media of randomly packed granules", *Applied Mechanics Reviews*, vol. 47, no. 1S, pp. S197–S206, 1994.

[CHA 05] CHANG C.S., HICHER P.Y., "An elasto-plastic model for granular materials with microstructural consideration", *International Journal of Solids and Structures*, vol. 42, no. 14, pp. 4258–4277, 2005.

[CHA 14] CHAZE M., CAMBOU B., "Analysis of internal state and strains in granular material at meso-scale: influence of particle shape", *Granular Matter*, vol. 16, no. 5, pp. 657–673, 2014.

[CHR 81] CHRISTOFFERSON J., MEHRABADI M.M., NEMAT NASSER S., "A micromechanical description of granular material behavior", *Journal of Applied Mechanics*, vol. 48, no. 2, pp. 339–344, 1981.

[CUN 79] CUNDALL P.A., STRACK O.D.L., "A discrete numerical model for granular assemblies", *Géotechnique*, vol. 29, no. 1, pp. 47–65, 1979.

[DED 00] DEDECKER F., CHAZE M., DUBUJET P.H. *et al.*, "Specific features of strain in granular materials", *Mechanics of Cohesive-Frictional Materials*, vol. 5, no. 3, pp. 173–193, 2000.

[DEL 90] DELYON F., DUFRESNE D., LÉVY Y., "Physique et génie civil, deux illustrations simples", *Annales des Ponts et Chaussées, numéro spécial: Mécanique des milieux granulaires*, pp. 22–29, 1990.

[DUB 06] DUBOIS F., JEAN M., "The non-smooth contact dynamic method: recent LMGC90 software developments and application", in WRIGGERS P., NACKENHORST U. (eds), *Analysis and Simulation of Contact Problems*, Springer Berlin Heidelberg, pp. 375–378, 2006.

[DUR 10a] DURÁN O., KRUYT N.P., LUDING S., "Micro-mechanical analysis of deformation characteristics of three-dimensional granular materials", *International Journal of Solids and Structures*, vol. 47, no. 17, pp. 2234–2245, 2010.

[DUR 10b] DURÁN O., KRUYT N.P., LUDING S., "Analysis of three-dimensional micro-mechanical strain formulations for granular materials: Evaluation of accuracy", *International Journal of Solids and Structures*, vol. 47, no. 2, pp. 251–260, 2010.

[DUR 15] DURIEZ J., VINCENS E., "Constitutive modelling of cohesionless soils and interfaces with various internal states: an elasto-plastic approach", *Computers and Geotechnics*, vol. 63, pp. 33–45, 2015.

[EDW 98] EDWARDS S.F., "The equations of stress in a granular material", *Physica A: Statistical Mechanics and its Applications*, vol. 249, nos. 1–4, pp. 226–231, 1998.

[EHL 03] EHLERS W., RAMM E., DIEBELS S. *et al.*, "From particle ensembles to Cosserat continua: homogenization of contact forces towards stresses and couple stresses", *International Journal of Solids and Structures*, vol. 40, no. 24, pp. 6681–6702, 2003.

[EME 96] EMERIAULT F., CAMBOU B., "Micromechanical modelling of anisotropic nonlinear elasticity of granular medium", *International Journal of Solids and Structures*, vol. 33, no. 18, pp. 2591–2607, 1996.

[FOR 03] FORTIN J., MILLET O., DE SAXCÉ G., "Construction of an averaged stress tensor for a granular medium", *European Journal of Mechanics, A / Solids*, vol. 22, no. 4, pp. 567–582, 2003.

[GOL 02] GOLDHIRSCH I., GOLDENBERG C., "On the microscopic foundations of elasticity", *European Physical Journal E*, vol. 9, no. 3, pp. 245–251, 2002.

[GUO 07] GUO P., SU X., "Shear strength, interparticle locking, and dilatancy of granular materials", *Canadian Geotechnical Journal*, vol. 44, no. 5, pp. 579–591, 2007.

[ITA 99] ITASCA INC., PFC2D – theory and background, *Itasca Consulting Group, Inc.*, 1999.

[JEA 99] JEAN M., "The non-smooth contact dynamics method", *Computer Methods in Applied Mechanics and Engineering*, vol. 177, nos. 3–4, pp. 235–257, 1999.

[JEN 91] JENKINS J.T., "Anisotropic elasticity for random arrays of identical spheres", in WU J. (ed.), *Modern Theory of Anisotropic Elasticity and Applications*, Society for Industrial Applied Mathematics, pp. 368–377, 1991.

[KON 82] KONISHI J., ODA M., NEMAT-NASSER S., "Inherent anisotropy and shear strength of assembly of oval cross-sectional rods," in *Proceedings of the Deformation and Failure of Granular Materials, International Union of Theoretical and Applied Mechanics Symposium*, A. A. Balkema, Rotterdam, pp. 403-412, 1982.

[KRU 96] KRUYT N.P., ROTHENBURG L., "Micromechanical definition of the strain tensor for granular materials", *Journal of Applied Mechanics*, vol. 63, no. 3, pp. 706–711, 1996.

[KRU 02] KRUYT N.P., ROTHENBURG L., "Probability density functions of contact forces for cohesionless frictional granular materials", *International Journal of Solids and Structures*, vol. 39, no. 3, pp. 571–583, 2002.

[KRU 14] KRUYT N.P., MILLET O., NICOT F., "Macroscopic strains in granular materials accounting for grain rotations", *Granular Matter*, vol. 16, no. 6, pp. 933–944, 2014.

[KUH 97] KUHN M.R., "Deformation measures for granular materials", in CHANG C.S. (ed.), *Mechanics of Deformation and Flow of Particulate Materials*, ASCE, pp. 91–104, 1997.

[KUH 99] KUHN M.R., "Structured deformation in granular materials", *Mechanics of Materials*, vol. 31, no. 6, pp. 407–429, 1999.

[KUH 03] KUHN M.R., "Heterogeneity and patterning in the quasi-static behavior of granular materials", *Granular Matter*, vol. 4, no. 4, pp. 155–166, 2003.

[LÄT 00] LÄTZEL M., LUDING S., HERRMANN H.J., "Macroscopic material properties from quasi-static, microscopic simulations of a two-dimensional shear-cell", *Granular Matter*, vol. 2, no. 3, pp. 123–135, 2000.

[LIA 97] LIAO C.L., CHANG T.P., YOUNG D.H. *et al.*, "Stress strain relationship for granular materials based on the hypothesis of best fit", *International Journal of Solids and Structures*, vol. 34, nos. 31–32, pp. 4087–4100, 1997.

[LOV 44] LOVE A.E.H., *A Treatise on Mathematical Theory of Elasticity*, Dover Publications, New York, 1944.

[LUD 08] LUDING S., "Introduction to discrete element methods: basic of contact force models and how to perform the micro-macro transition to continuum theory", *European Journal of Environmental and Civil Engineering*, vol. 12, nos. 7–8, pp. 785–826, 2008.

[MAG 08] MAGOARIEC H., DANESCU A., CAMBOU B., "Non-local orientational distribution of contact forces in granular samples containing elongated particles", *Acta Geotechnica*, vol. 3, no. 1, pp. 49–60, 2008.

[MAL 00] MALEKI M., DUBUJET P., CAMBOU B., "Modélisation hiérarchisée du comportement des sols", *Revue Française de génie civil*, vol. 4, nos. 7–8, pp. 895–928, 2000.

[MOR 93] MOREAU J.J., "New computation methods in granular dynamics", in THORNTON C. (ed.), *Powders & Grains 93*, CRC Press, 1993.

[MOR 97] MOREAU J.J., "Numerical investigation of shear zones in granular materials", in GRASSBERGER P., WOLFS D. (eds), *Friction, Arching, Contact Dynamics*, World Scientific, Singapore, pp. 233–247, 1997.

[MOR 10] MOREAU J.J., "The stress tensor in granular media and in other mechanical collections", in CAMBOU B., JEAN M., RADJAÏ F. (eds), *Micromechanics of Granular Materials*, pp. 51–100, ISTE Ltd, London and John Wiley Sons, New York, 2010.

[NEM 83] NEMAT-NASSER S., MEHRABADI M., "Stress and fabric in granular masses", in JENKINS J.T., SATAKE M. (eds), *Mechanics of Granular Materials: New Models and Constitutive Relations*, Elsevier Science, Amsterdam, pp. 1–8, 1983.

[NGU 09a] NGUYEN N.S., Prise en compte d'une échelle mésoscopique dans l'étude du comportement des milieux granulaires, PhD thesis, Ecole Centrale de Lyon, France, 2009.

[NGU 09b] NGUYEN N.S., MAGOARIEC H., CAMBOU B. *et al.*, "Analysis of structure and strain at the meso-scale in 2D granular materials", *International Journal of Solids and Structures*, vol. 46, no. 17, pp. 3257–3271, 2009.

[NGU 12] NGUYEN N.S., MAGOARIEC H., CAMBOU B., "Local stress analysis in granular materials at a mesoscale", *International Journal for Numerical and Analytical Methods in Geomechanics*, vol. 36, no. 14, pp. 1609–1635, 2012.

[NGU 14] NGUYEN N.S., MAGOARIEC H., CAMBOU B., "Analysis of local behavior in granular materials", *Comptes Rendus – Mécanique*, vol. 342, no. 3, pp. 156–173, 2014.

[NGU 15] NGUYEN S.K., MAGOARIEC H., VINCENS E. et al., "Towards a new approach for modeling the behavior of granular materials: a mesoscopic-macroscopic change of scale", *International Journal of Solids and Structures*, pp. 1–40, 2015.

[NIC 11] NICOT F., DARVE F., "The H-microdirectional model: accounting for a mesoscopic scale", *Mechanics of Materials*, vol. 43, no. 12, pp. 918–929, 2011.

[NOU 05] NOUGUIER-LEHON C., VINCENS E., CAMBOU B., "Structural changes in granular materials: the case of irregular polygonal particles", *International Journal of Solids and Structures*, vol. 42, nos. 24–25, pp. 6356–6375, 2005.

[ODA 72] ODA M., "The mechanism of fabric change during compressional deformation of sands", *Soils and Foundations*, vol. 12, no. 2, pp. 1–18, 1972.

[ODA 82] ODA M., KONISHI M., NEMAT-NASSER S., "Experimental micromechanical evaluation of strength of granular materials: effects of particle rolling", *Mechanics of Materials*, vol. 1, no. 4, pp. 269–283, 1982.

[OSU 11] O'SULLIVAN C., "Particle-based discrete element modeling: geomechanics perspective", *International Journal of Geomechanics*, vol. 11, no. 6, pp. 449–464, 2011.

[RAD 98] RADJAÏ F., WOLF D.E., JEAN M. et al., "Bimodal character of stress transmission in granular packings", *Physical Review Letters*, vol. 80, no. 1, pp. 61–64, 1998.

[RAD 09] RADJAÏ F., RICHEFEU V., "Contact dynamics as a non-smooth discrete element method", *Mechanics of Materials*, vol. 41, no. 6, pp. 715–728, 2009.

[REB 08] REBOUL N., VINCENS E., CAMBOU B., "A statistical analysis of void size distribution in a simulated narrowly graded packing of spheres", *Granular Matter*, vol. 10, no. 6, pp. 457–468, 2008.

[ROT 81] ROTHENBURG L., SELVADURAI A.S., "Micromechanical aspects of random assemblies of spheres with linear contact interactions", *Eighth Canadian Congress of Applied Mechanics*, pp. 217–218, 1981.

[ROU 02] ROUX J.N., COMBE G., "Quasistatic rheology and the origins of strain", *Comptes Rendus Physique*, vol. 3, no. 2, pp. 131–140, 2002.

[SAB 97] SAB K., "A new approach of homogenization for granular media", *Saint-Venant Symposium: Analyse multiéchelle et systèmes physiques couplés*, Presses de l'Ecole Nationale des Ponts et Chausées, Paris, pp. 597–603, 1997.

[SAT 78] SATAKE M., "Constitution of mechanics of granular materials through the graph theory", *Continuum Mechanical and Statistical Approaches in the Mechanics of Granular Materials*, pp. 47–62, 1978.

[SAT 82] SATAKE M., "Fabric tensor in granular materials", *IUTAM conference on deformation and failure of granular materials, Delft*, pp. 63–68, 1982.

[SAT 92] SATAKE M., "A discrete-mechanical approach to granular materials", *International Journal of Engineering Science*, vol. 30, no. 10, pp. 1525–1533, 1992.

[SAT 04] SATAKE M., "Tensorial form definitions of discrete-mechanical quantities for granular assemblies", *International Journal of Solids and Structures*, vol. 41, no. 21, pp. 5775–5791, 2004.

[TAM 05] TAMAGNINI C., CALVETTI F., VIGGIANI G., "An assessment of plasticity theories for modeling the incrementally nonlinear behavior of granular soils", *Journal of Engineering Mathematics*, vol. 52, nos. 1–3, pp. 265–291, 2005.

[TAT 92] TATSUOKA F., SHIBUYA S., "Deformation characteristics of soils and rocks from field and laboratory tests", *The 9th Asian Regional Conference on SMFE*, Bangkok, vol. II, pp. 101–170, 1992.

[TAY 48] TAYLOR D.W., *Fundamentals of Soil Mechanics*, John Wiley & Sons, 1948.

[THO 00] THORNTON C., "Numerical simulations of deviatoric shear deformation of granular media", *Géotechnique*, vol. 50, no. 1, pp. 43–53, 2000.

[TSU 98] TSUCHIKURA T., SATAKE M., "Statistical measure tensors and their application to computer simulation analysis of biaxial compression test", in MURAKAMI H., LUCO J.E. (eds), *Engineering Mechanics: A Force for 21st Century*, ASCE, pp. 1732–1735, 1998.

[WAL 87] WALTON K., "The effective elastic moduli of a random packing of spheres", *Journal of the Mechanics and Physics of Solids*, vol. 35, no. 2, pp. 213–226, 1987.

[WAN 70] WANG C.C., "A new representation theorem for isotropic functions: an answer to Professor G.F. Smith's criticism of my papers on representation of isotropic functions", *Archive for Rational Mechanics and Analysis*, vol. 36, no. 3, pp. 166–223, 1970.

[WEB 66] WEBER J., "Recherches concernant les contraintes intergranulaires dans les milieux pulvrulents", *Bulletin de liaison des ponts et chaussées*, vol. 20, pp. 1–20, 1966.

[ZHU 16] ZHU H., NICOT F., DARVE F., "Meso-structure evolution in a 2D granular material during biaxial loading", *Granular Matter*, vol. 18, no. 1, pp. 1–12, 2016.

Index

Printed in the United States
By Bookmasters